SOURCE CODING THEORY

THE KLUWER INTERNATIONAL SERIES IN ENGINEERING AND COMPUTER SCIENCE

COMMUNICATIONS AND INFORMATION THEORY

Consulting Editor

Robert Gallager

Other books in the series:

Digital Communications. Edward A. Lee, David G. Messerschmitt
ISBN 0-89838-274-2

An Introduction to Cryptology. Henk C. A. van Tilborg
ISBN 0-89838-271-8

Finite Fields for Computer Scientists and Engineers. Robert J. McEliece
ISBN 0-89838-191-6

An Introduction to Error Correcting Codes With Applications Scott A. Vanstone and Paul C. van Oorschot, ISBN 0-7923-9017-2

SOURCE CODING THEORY

by

Robert M. Gray

Stanford University

KLUWER ACADEMIC PUBLISHERS
Boston/Dordrecht/London

Distributors for North America:
Kluwer Academic Publishers
101 Philip Drive
Assinippi Park
Norwell, Massachusetts 02061 USA

Distributors for all other countries:
Kluwer Academic Publishers Group
Distribution Centre
Post Office Box 322
3300 AH Dordrecht, THE NETHERLANDS

Library of Congress Cataloging-in-Publication Data

Gray, Robert M., 1943-
Source coding theory / by Robert Gray.
p. cm. — (The Kluwer international series in engineering and computer science. Communications and information theory)
Includes bibliographical references.
ISBN 0-7923-9048-2
1. Coding theory. 2. Rate distortion theory. I. Title.
II. Series.
TK5102.5.G69 1990
003'.54—dc20 89-39016
CIP

Printed in the United States of America.

This printing is a digital duplication of the original edition.

à mes amies

Contents

Preface

Source coding theory has as its goal the characterization of the optimal performance achievable in idealized communication systems which must code an information source for transmission over a digital communication or storage channel for transmission to a user. The user must decode the information into a form that is a good approximation to the original. A code is *optimal* within some class if it achieves the best possible fidelity given whatever constraints are imposed on the code by the available channel. In theory, the primary constraint imposed on a code by the channel is its rate or resolution, the number of bits per second or per input symbol that it can transmit from sender to receiver. In the real world, complexity may be as important as rate.

The origins and the basic form of much of the theory date from Shannon's classical development of noiseless source coding and source coding subject to a fidelity criterion (also called rate-distortion theory) [73] [74]. Shannon combined a probabilistic notion of information with limit theorems from ergodic theory and a random coding technique to describe the optimal performance of systems with a constrained rate but with unconstrained complexity and delay. An alternative approach called asymptotic or high rate quantization theory based on different techniques and approximations was introduced by Bennett at approximately the same time [4]. This approach constrained the delay but allowed the rate to grow large.

The goal of both approaches was to provide unbeatable bounds to the achievable performance using realistic code structures on reasonable mathematical models of real-world source coding systems such as analog-to-digital conversion, data compression, and entropy coding. The original theory dealt almost exclusively with a particular form of code—a block code or, as it is sometimes called in current applications, a vector quantizer. Such codes operate on nonoverlapping blocks or vectors of input symbols in a memoryless fashion, that is, in a way that does not depend on previous blocks. Much of the theory also concentrated on memoryless sources or sources with very simple memory structure. These results have since been extended to a variety of coding structures and to far more general sources. Unfortunately, however, most of the results for nonblock codes have not appeared in book form and their proofs have involved a heavy dose of measure theory and ergodic theory. The results for nonmemoryless sources have also usually been either difficult to prove or confined to Gaussian sources.

This monograph is intended to provide a survey of the Shannon coding theorems for general sources and coding structures along with a treatment of high rate vector quantization theory. The two theories are compared and contrasted. As perhaps the most important special case of the theory, the uniform quantizer is analyzed in some detail and the behavior of quantization noise is compared and contrasted with that predicted by the theory and approximations. The treatment includes examples of uniform quantizers used inside feedback loops. In particular, the validity of the common white noise approximation is examined for both Sigma-Delta and Delta modulation. Lattice vector quantizers are also considered briefly.

Much of this manuscript was originally intended to be part of a book by Allen Gersho and myself titled *Vector Quantization and Signal Compression* which was originally intended to treat in detail both the design algorithms and performance theory of source coding. The project grew too large, however, and the design and applications-oriented material eventually crowded out the theory. This volume can be considered as a theoretical companion to *Vector Quantization and Signal Compression*, which will also be published by Kluwer Academic Press.

Although a prerequisite graduate engineering level of mathematical sophistication is assumed, this is not a mathematics text and I have been admittedly somewhat cavalier with the mathematical details. The arguments always have a solid foundation, however, and the interested reader can pursue them in the literature. In particular, a far more careful and detailed treatment of most of these topics may be found in my manuscript *Mathematical Information Theory*.

The principal existing text devoted to source coding theory is Berger's book on rate-distortion theory [5]. Gallager's chapter on source coding in [28] also contains a thorough and oft referred-to treatment. Topics treated here that are either little treated or not treated at all in Berger and Gallager include sliding-block codes, feedback and finite-state codes, trellis encoders, synchronization of block codes, high rate vector quantization theory, process definitions of rate-distortion functions, uniform scalar quantizer noise theory, and Sigma-Delta and Delta modulation noise theory in scalar quantization (or PCM), in Sigma-Delta modulation, and in Delta modulation. Here the basic source coding theorem for block codes is proved without recourse to the Nedoma decomposition used by Berger and Gallager [5], [28]. The variational equations defining the rate-distortion function are developed using Blahut's approach [6], but both Gallager's and Berger's solutions are provided as each has its uses. The Blahut and Gallager/Berger approaches are presented in some detail and the proofs given take advantage of both viewpoints. In particular, the use of calculus optimizations is minimized by repeated applications of the divergence inequality.

The primary topic treated by Berger and Gallager and not included here is that of continuous time source coding theory. Noiseless coding is also not treated here as it is developed in the companion volume as well as in most information theory texts (as well as texts devoted entirely to noiseless coding, e.g., [77].)

This book is intended to be a short but complete survey of source coding theory, including rate-distortion theory, high rate quantization theory, and uniform quantizer noise theory. It can be used in conjunction with *Vector Quantization and Signal Compression* in a graduate level course to provide either background reading on the underlying theory or as a supplementary text if the course devotes time to both the theory and the design of vector quantizers. When I teach an "Advanced Topics" course on data compression, I proceed from a brief review of the prerequisite random process material into Part III of *Vector Quantization and Signal Compression*, that is, directly into the development of code design algorithms for vector quantizers. I treat the design material first as this is usually the preferred material for term papers or projects. The second half of the course is then devoted to Chapters 3, 4, and 5 of this book. Portions of Chapter 6 are treated if time permits and are used to point out the shortcomings of the asymptotic approximations as well as to provide an introduction into the theory of oversampled analog-to-digital converters.

Robert M. Gray
Stanford, California
July 1989

Acknowledgements

The research in information theory that yielded many of the results and some of the new proofs for old results in this book was supported by the National Science Foundation. The research on the theory of Sigma-Delta modulation quantizer noise was supported by a seed research grant from the Stanford University Center for Integrated Systems and by the National Science Foundation. The book was originally written using the eqn and troff utilities and was subsequently translated into LaTeX on several UNIX (an AT&T trademark) systems and on Apple Macinstosh-SEs supported by the Industrial Affiliates Program of the Stanford University Information Systems Laboratory. The book benefited greatly from numerous comments by Ender Ayanoğlu, Phil Chou, Tom Lookabaugh, Nader Moayeri, Eve Riskin, Maurizio Longo, Wu Chou, and Sang Ju Park as well as from numerous corrections and improvements by graduate students.

SOURCE CODING THEORY

Chapter 1

Information Sources

An information source is modeled mathematically as a discrete time random process, a sequence of random variables. This permits the use of all of the tools of the theory of probability and random processes. In particular, it allows us to do theory with probabilistic averages or expectations and relate these to actual time average performance through laws of large numbers or ergodic theorems. Such theorems describing the long term behavior of well behaved random systems are crucial to such theoretical analysis. Relating expectations and long term time averages requires an understanding of stationarity and ergodic properties of random processes, properties which are somewhat difficult to define precisely, but which usually have a simple intuitive interpretation. These issues are not simply of concern to mathematical dilettants. For example, stationarity can be violated by such commonly occurring phenomena as transients and variable length coding, yet sample averages may still converge in a useful way. In this chapter we survey some of the key ideas from the theory of random processes. The chapter strongly overlaps portions Chapter 2 of Gersho and Gray [30] and is intended to provide the necessary prerequisites and establish notation. The reader is assumed to be familiar with the general topics of probability and random processes and the chapter is intended primarily for reference and review. A more extensive treatment of basic random processes from a similar point of view may be found in [40] and [37].

1.1 Probability Spaces

The basic building block of the theory of probability and random processes is the *probability space* or *experiment*, a collection of definitions and axioms which yield the calculus of probability and the basic limit theorems relating expectations and sample averages. For completeness we include the basic definitions and examples. A *probability space* $(\Omega, \mathcal{F}, P)$ is a collection of three things:

Ω An abstract space Ω called the *sample space*. Intuitively this is a listing of all conceivable outcomes of the experiment.

$\mathcal{F}$ A nonempty collection of subsets of Ω called *events* which has the following properties:

1. If $F \in \mathcal{F}$, then also $F^c \in \mathcal{F}$, that is, if the set F is in the collection $\mathcal{F}$, then so is its complement $F^c = \{\omega : \omega \notin F\}$.
2. If $F_i \in \mathcal{F}$, $i = 1, 2, \ldots$, then also $\bigcup_i F_i \in \mathcal{F}$, that is, if a (possibly infinite) collection of sets F_i belong to the collection $\mathcal{F}$, then so does the union $\bigcup_i F_i \in \mathcal{F} = \{\omega : \omega \in F_i \text{ for some } i\}$.

It follows from the above conditions that $\Omega \in \mathcal{F}$, i.e., the set "something happens" is an event, and the null set $\emptyset = \Omega^c$ ("nothing happens") is an event. A collection $\mathcal{F}$ of subsets of Ω with these properties is called an *event space* or a *σ-field.* It is a standard result that the above two conditions imply that any sequence of complements, unions, or intersections of events (members of $\mathcal{F}$) yields another event. This provides a useful algebraic structure to the collection of sets for which we wish to define a probability measure.

P A *probability measure* on an event space $\mathcal{F}$ of subsets of a sample space Ω is an assignment of a real number $P(F)$ to every F in $\mathcal{F}$ which satisfies the following rules (often called the *axioms of probability*):

1. $P(F) \geq 0$ for all $F \in \mathcal{F}$, that is, probabilities are nonnegative.
2. $P(\Omega) = 1$, that is, probability of the entire sample space ("something happens") is 1.
3. If events F_i, $i = 1, 2, \ldots$ are disjoint, that is, if $F_i \bigcap F_j = \{\omega : \omega \in F_i \text{ and } \omega \in F_j\} = \emptyset$ for all $i \neq j$, then

$$P(\bigcup_{i=1}^{\infty} F_i) = \sum_{i=1}^{\infty} P(F_i);$$

that is, the probability of the union of a sequence of disjoint events is the sum of the probabilities.

It is important to note that probabilities need be defined only for events and not for *all* subsets of the sample space. The abstract setup of a probability space becomes more concrete if we consider the most important special cases. One basic example is the case where $\Omega = \mathcal{R} = (-\infty, \infty)$, the real line. The most useful event space in this case is called the *Borel field* and is denoted $\mathcal{B}(\mathcal{R})$. We will not delve into the theoretical details of Borel fields, but we shall point out that it contains all of the subsets of the real line that are sufficiently "nice" to have consistent probabilities defined on them, e.g., all intervals and sets constructable by sequences of complements, unions, and intersections of intervals. It is a basic result of measure theory, however, that the Borel field does not contain *all* subsets of the real line. The members of the Borel field, that is, the events in the real line, are called *Borel sets*.

Suppose that we have a real valued function f defined on $\mathcal{R}$ with the following properties:

1. $f(r) \geq 0$, all $r \in \mathcal{R}$,

2. $\int_{-\infty}^{\infty} f(r)\, dr = 1$.

Then the set function P defined by

$$P(F) = \int_F f(r)\, dr$$

is a probability measure and the function f is called a *probability density function* or *pdf* since it is integrated to find probability. There is more going on here than meets the eye and a few words of explanation are in order. Strictly speaking, the above claim is true only if the integral is considered as a Lebesgue integral rather than as the Riemann integral familiar to most engineers. As is discussed in some length in [40], however, this can be considered as a technical detail and Riemann calculus can be used without concern provided that the Riemann integrals make sense, that is, can be evaluated. If the Riemann integral is not well defined, then appropriate limits must be considered. Some of the more common pdf's are listed below. The pdf's are 0 outside the listed domain. $b > a$, $\lambda > 0$, m, and $\sigma > 0$ are real-valued parameters which specify the pdf's.

The uniform pdf $f(r) = 1/(b-a)$ for $r \in [a, b] = \{r : a \leq r \leq b\}$.

The exponential pdf $f(r) = \lambda e^{-\lambda r}$; $r \geq 0$.

The doubly exponential or Laplacian pdf $f(r) = \frac{\lambda}{2} e^{-\lambda |r|}$; $r \in \mathcal{R}$.

The Gaussian pdf $f(r) = (2\pi\sigma^2)^{-1/2} e^{-(r-m)^2/2\sigma^2}$; $r \in \mathcal{R}$.

A similar construction works when Ω is some subset of the real line instead of the entire real line. In that case the appropriate event space comprises all the Borel sets in Ω. For example, we could define a probability measure on $([0,\infty),\mathcal{B}([0,\infty)))$, where $\mathcal{B}([0,\infty))$ denotes the Borel sets in $[0,\infty)$, using the exponential pdf. Obviously we could either define this experiment on the entire real line using a pdf that is 0 for negative numbers and exponential on nonnegative numbers or we could define it on the nonnegative portion of the real line using a pdf that is an exponential everywhere. This is strictly a matter of convenience. Another common construction of a probability measure arises when all of the probability sits on a discrete subset of the real line (or any other sample space). Suppose that Ω' is an arbitrary sample space and $\mathcal{F}$ a corresponding event space. Suppose that an event $\Omega \subset \Omega'$ consists of a finite or countably infinite collection of points. (By countably infinite we mean a set that can be put into one-to-one correspondence with the nonnegative integers, e.g., the nonnegative integers, the integers, the even integers, and the rational numbers.) Suppose further that we have a function p defined for all points in Ω which has the following properties:

1. $p(\omega) \geq 0$ all $\omega \in \Omega$.

2. $\sum_{\omega \in \Omega} p(\omega) = 1$.

Then the set function P defined by

$$P(F) = \sum_{\omega \in F \bigcap \Omega} p(\omega)$$

is a probability measure and the function p is called a *probability mass function* or *pmf* since one adds the probability masses of points to find the overall probability. That P defined as a sum over a pmf is indeed a probability measure follows from the properties of sums. (It is also a special case of the Lebesgue integral properties.) Some of the more common pmf's are listed below. The pmf's $p(\omega)$ are specified in terms of parameters: p is a real number in $(0,1)$, n is a positive integer, and λ is a positive real number.

The binary pmf $\Omega = \{0,1\}$. $p(1) = p$, $p(0) = 1 - p$.

The uniform pmf $\Omega = \{0,1,\ldots,n-1\}$. $p(k) = 1/n$; $k \in \Omega$.

The geometric pmf $\Omega = \{1,2,\ldots\}$. $p(k) = p(1-p)^{k-1}$; $k \in \Omega$.

The Poisson pmf $\Omega = \{0,1,2,\ldots\}$. $p(k) = \lambda^k e^{-\lambda}/k!$; $k \in \Omega$.

In summary, one can define probability measures on the real line by integrating pdf's or by summing pmf's. Note that probability measures are set functions; that is, they are defined for all events or sets in an event space. Probability functions such as pdf's and pmf's, however, are point functions: They are defined for all points in the sample space. The probability functions must satisfy properties similar to those of probability measures, i.e., nonnegativity and normalization. The countable additivity property of probability measures then follows from the properties of integration or summation. One can also combine pdf's and pmf's into a single probability measure; e.g., suppose that P_1 is a probability measure on Ω' defined by a pdf f and P_2 is a probability measure defined by a pmf p such that $P_2(\Omega) = 1$ for some discrete set $\Omega \subset \Omega'$. Given any $\lambda \in [0,1]$ we can define a *mixture probability measure* P by

$$\begin{aligned} P(F) &= \lambda P_1(F) + (1-\lambda)P_2(F) \\ &= \lambda \int_F f(r)\,dr + (1-\lambda) \sum_{r \in F \cap \Omega} p(r). \end{aligned}$$

P can represent an experiment with both continuous and discrete components. By choosing λ to be 1 (or 0) we have the purely continuous (or discrete) special cases. It is common practice and sometimes convenient to use a pdf for such a mixed probability distribution using Dirac delta functions. With this notation, the composite pdf $f_c(r)$ can be written as

$$f_c(r) = \lambda f(r) + (1-\lambda) \sum_{s \in \Omega} p(s)\delta(r-s)$$

which defines the same mixed probability measure as above assuming the same care is taken in defining the integral of the pdf over a set with a boundary point located at a mass point in F. One might suspect that such a mixture of discrete and continuous parts is the most general form of probability measure on the real line, but this is unfortunately not the case. There is a third type probability measure that is called *singular* which cannot be evaluated as an integral of a pdf or a sum of a pmf. Happily one rarely encounters such probability measures and hence we can usually consider the above mixtures to be the most general form.

Another probability function of interest that is defined in both the continuous and discrete case is the *cumulative distribution function* or *cdf* as follows: Given a probability measure P on the real line (or some subset thereof), define the cdf $F(r)$ by

$$F(r) = P(\{\omega : \omega \leq r\});$$

that is, $F(r)$ is simply the probability of getting a value less than or equal to r. If P is defined by a pdf f, then

$$F(r) = \int_{-\infty}^{r} f(x)\,dx$$

and hence the fundamental theorem of calculus implies that

$$f(r) = \frac{dF(r)}{dr},$$

and hence the pdf is the derivative of the cdf. Alternatively, if the cdf is differentiable, then the pdf exists and the probability measure of events can be found by integrating the pdf. In the discrete case, the pmf can be found as the differences of the cdf, e.g., if Ω is the set of integers, then $p(k) = F(k) - F(k-1)$.

The above ideas are easily extended to sample spaces consisting of real valued vectors rather than just real numbers, that is, Euclidean vector spaces. Suppose that $\Omega = \mathcal{R}^n$, the space of all vectors $\mathbf{x} = (x_0, x_1, \ldots, x_{n-1})$ with $x_i \in \mathcal{R}$. A Borel field can be constructed for this sample space, but we will not consider the details. Suffice it to say that it contains all sets usually encountered, e.g., multidimensional rectangles and all sets constructable by sequences of set theoretic operations on rectangles.

A function f defined on $\mathcal{R}^n$ is called a *joint pdf* or *n-dimensional pdf* or simply a *pdf* if

1. $f(\mathbf{x}) \geq 0,\ \mathbf{x} \in \mathcal{R}^n$.
2. $\int_{\mathbf{x} \in \mathcal{R}^n} f(\mathbf{x})\,d\mathbf{x} = 1$,
   where the vector integral notation is shorthand for

$$\int_{x_0 \in \mathcal{R}} \int_{x_1 \in \mathcal{R}} \cdots \int_{x_{n-1} \in \mathcal{R}} f(x_0, x_1, \ldots, x_{n-1})\,dx_0\,dx_1 \ldots dx_{n-1}.$$

Then the set function P defined by the integral

$$P(G) = \int_G f(\mathbf{x})\,d\mathbf{x}$$

is a probability measure (subject to considerations like those of the scalar case).

In a similar manner we can consider pmf's for vectors. We can completely describe a probability measure on Euclidean space by giving the cdf (or in special cases a pdf or pmf). This lets us describe two of the most common and important examples of probability measures on vector spaces: The product measure and the Gaussian measure. Suppose that we have a collection of cdf's $F_i(r)$; $i = 0, \ldots, n-1$. They may be identical or different. Define an n-dimensional cdf $F^n(\mathbf{r})$ by

$$F^n(\mathbf{r}) = \prod_{i=0}^{n-1} F_i(r_i).$$

(It is straightforward to verify that F^n is indeed a cdf.) A cdf of this form is called a *product cdf* and the resulting probability measure is called a product probability measure. If the original cdf's F_i are defined by densities f_i, then the above is equivalent to saying that the n-dimensional pdf f^n is given by

$$f^n(\mathbf{r}) = \prod_{i=0}^{n-1} f_i(r_i).$$

Similarly, if the original cdf's F_i are defined by pmf's p_i, then the above is equivalent to saying that the n-dimensional pmf p^n is given by

$$p^n(\mathbf{r}) = \prod_{i=0}^{n-1} p_i(r_i).$$

Thus any of the scalar pdf's or pmf's previously defined can be used to construct product measures on vector spaces. The second important example is the multidimensional Gaussian pdf. Here we describe the n-dimensional pdf which in turn implies the probability measure and the cdf. A pdf f on $\mathcal{R}^n$ is said to be *Gaussian* if it has the following form:

$$f(\mathbf{x}) = \frac{e^{-\frac{1}{2}(\mathbf{x}-\mathbf{m})^t \Lambda^{-1}(\mathbf{x}-\mathbf{m})}}{\sqrt{(2\pi)^n \det \Lambda}},$$

where the superscript t stands for "transpose", $\mathbf{m} = (m_0, \ldots, m_{n-1})^t$ is a column vector, and Λ is an $n \times n$ square symmetric matrix, e.g., $\Lambda = \{\Lambda(i,j);\ i = 0, \ldots, n-1;\ j = 0, \ldots, n-1\}$ and $\Lambda(i,j) = \Lambda(j,i)$, all i, j. Equivalently, $\Lambda^t = \Lambda$. We also require that Λ be strictly positive definite, that is, for any nonzero vector $\mathbf{y}$ the quadratic form $\mathbf{y}^t \Lambda \mathbf{y}$ is positive, that is,

$$\mathbf{y}^t \Lambda \mathbf{y} = \sum_{i=0}^{n-1} \sum_{j=0}^{n-1} y_i \Lambda_{i,j} y_j > 0.$$

A standard result of matrix theory ensures that if Λ is strictly positive definite, then the inverse exists and the determinant is not 0 and hence the above pdf is well defined.

1.2 Random Variables and Vectors

A random variable can be thought of as a measurement on a probability space: an experiment produces some outcome ω and an observer then measures or some function of ω, say $X(\omega)$. This measurement a random variable defined on the underlying probability space or experiment. For example, if ω is a real vector, X could be its Euclidean norm, the minimum coordinate value, the third component, and so on. The basic idea is that if ω is produced in a random fashion as described by a probability space, then X should inherit a probabilistic description from the underlying space, thereby giving us a new probability space which describes the possible outcomes of X. Suppose then that we have a probability space $(\Omega, \mathcal{F}, P)$ and a function or mapping $X : \Omega \to \mathcal{R}$, that is, X assigns a real number to every point in the sample space. We wish to construct a new probability space that describes the outcomes of X, say $(\mathcal{R}, \mathcal{B}(\mathcal{R}), P_X)$, where $P_X(G)$ is the probability that the measurement X assumes a value in the event G. We abbreviate this English statement to $\Pr(X \in G)$. We should be able to compute this probability by referring back to the original probability space, that is, we should have that $P_X(G) = \Pr(X \in G) = P(\omega : \; X(\omega) \in G) = P(X^{-1}(G))$, where the *inverse image* $X^{-1}(G)$ of a set G under the mapping X is defined is the collection of all ω points which map into G:

$$X^{-1}(G) = \{\omega : \; X(\omega) \in G\}.$$

In other words, $P_X(G)$ should be the probability of the set of sample points that are mapped by X into G. This will be the case *if and only if* the set $X^{-1}(G)$ is indeed an event in the original space, that is, a member of $\mathcal{F}$. We force this technical condition to be satisfied by building it into the definition of a random variable: Given a probability space $(\Omega, \mathcal{F}, P)$, a *random variable* X is a mapping from Ω into the real line $\mathcal{R}$ with the property that $X^{-1}(G) \in \mathcal{F}$ for all $G \in \mathcal{B}(\mathcal{R})$. Most reasonable functions encountered in engineering have this property, e.g., if the original space is a Euclidean vector space, then continuous functions are random variables and the maximum or minimum of the coordinates are random variables. The *distribution* of the random variable P_X is the induced probability measure defined by

$$P_X(G) = P(X^{-1}(G)); \; G \in \mathcal{B}(\mathcal{R}).$$

Thus, for example, if the probability measure on the original space is described by a pdf f, then

$$P_X(G) = \int_{r: X(r) \in G} f(r)\, dr.$$

If the probability measure on the original space is described by a pmf p, then analogously

$$P_X(G) = \sum_{k:\, X(k) \in G} p(k)$$

The distribution P_X is specified by a cdf F_X, which in turn may be described by a pmf p_X or a pdf f_X, that is,

$$P_X(G) = \int_{x \in G} f_X(x)\, dx$$

or

$$P_X(G) = \sum_{x \in G} p_X(x)$$

depending on whether X has a differentiable cdf or takes on discrete values.

Thus a probability space $(\Omega, \mathcal{F}, P)$ and a random variable X together induce a new probability space $(\mathcal{R}, \mathcal{B}(\mathcal{R}), P_X)$. If we begin with $\Omega = \mathcal{R}$, then a trivial random variable is $X(r) = r$, the identity function. In this manner all the pdf's and pmf's described for the real line can be considered as equivalent to distributions of random variables and the resulting random variables share the name of the probability functions. For example, we speak of a Gaussian random variable as one with a distribution induced by the Gaussian pdf. If the distribution P_X of a random variable X is induced by a pdf f or a pmf p, then we often subscript the probability function by the name of the random variable, i.e., f_X or p_X.

A *random vector* is simply a finite collection of random variables. For example, if X_i; $i = 0, 1, \ldots, n-1$, are random variables defined on a common probability space $(\Omega, \mathcal{F}, P)$, then $\mathbf{X} = (X_0, \ldots, X_{n-1})$ is a random vector. The *distribution* $P_{\mathbf{X}}$ of the random vector is the induced probability measure, that is,

$$P_{\mathbf{X}}(G) = P(\omega : \mathbf{X}(\omega) = (X_0(\omega), \ldots, X_{n-1}(\omega)) \in G) = P(\mathbf{X}^{-1}(G)).$$

It is a result of measure theory that the distribution is well defined.

As with any probability measure, a distribution can be specified by a cdf $F_{\mathbf{X}}(\mathbf{x}) = \Pr(X_i \leq x_i;\ i = 0, 1, \ldots, n-1)$ and possibly by a multidimensional pdf $f_{\mathbf{X}}$ or a pmf $p_{\mathbf{X}}$. As in the scalar case, we can define a trivial random vector on n-dimensional Euclidean space by $\mathbf{X}(\mathbf{r}) = \mathbf{r}$. If the probability measure is specified by a Gaussian (or other named) cdf, then the random vector is also said to be Gaussian (or whatever).

Given a probability space $(\Omega, \mathcal{F}, P)$, a finite collection of events F_i; $i = 0, \ldots, n-1$ is said to be *independent* if for any $k \leq n$ and sub-collection of events F_{l_i}; $i = 0, \ldots, k-1$ (where the l_i are distinct and $0 \leq l_i \leq n-1$) we have that

$$P(\bigcap_{i=0}^{k-1} F_{l_i}) = \prod_{i=0}^{k-1} P(F_{l_i}).$$

A collection of random variables X_i; $i = 0, \ldots, n-1$ is said to be independent if for all collections of Borel sets G_i; $i = 0, \ldots, n-1$, the events $X_i^{-1}(G_i)$ are independent. The definition is equivalent to saying that the random variables are independent if and only if

$$\Pr(X_i \in G_i; i = 0, \ldots, n-1) = \prod_{i=0}^{n-1} P_{X_i}(G_i), \tag{1.2.1}$$

for all collections of Borel sets G_i. It is straightforward to show that this is true if and only if the n-dimensional cdf for $\mathbf{X} = (X_0, \ldots, X_{n-1})$ is a product cdf. This definition has an interpretation in terms of elementary conditional probability. Given two events F and G for which $P(G) > 0$, the conditional probability of F given G is defined by

$$P(F|G) = \frac{P(F \bigcap G)}{P(G)}.$$

Thus if two events F and G are independent and $P(G) > 0$, then $P(F|G) = P(F)$; that is, the knowledge of G does not influence the probability of F. A random vector has independent components if and only if the appropriate probability functions factor in a manner analogous to 1.2.1, that is, if

$$F_{\mathbf{X}}(\mathbf{x}) = \prod_{i=0}^{n-1} F_{X_i}(x_i),$$

or

$$f_{\mathbf{X}}(\mathbf{x}) = \prod_{i=0}^{n-1} f_{X_i}(x_i),$$

or

$$p_{\mathbf{X}}(\mathbf{x}) = \prod_{i=0}^{n-1} p_{X_i}(x_i).$$

A random vector is said to be independent and identically distributed or *iid* if it has independent components with identical marginals, that is, all of the distributions P_{X_i} (or the corresponding probability functions) are identical.

1.3 Random Processes

A *random process* is an indexed family of random variables $\{X_t; t \in T\}$ or $\{X(t); t \in T\}$. The index t corresponds to time. Both forms of notation are used, but X_t is more common for discrete time and $X(t)$ more common for continuous time. If T is continuous, e.g., is $(-\infty, \infty)$ or $[0, \infty)$, then the process is called a *continuous time random process*. If T is discrete, as in the set of all integers or the set of all nonnegative integers, then the process is called a *discrete time random process* or a *random sequence* or a *time series*. When considering the discrete time case, letters such as n and k are commonly used for the index. The $t \in T$ is often omitted when it is clear from context and a random process denoted simply by $\{X_n\}$ or $\{X(t)\}$.

The above definition for a random process is abstract and is not always useful for actually constructing a mathematical model for a real process. The chief tool for this purpose is the Kolmogorov extension theorem, which we now describe. First, given a random process we can in principle compute the distribution for any finite dimensional random vectors obtainable by collecting a finite number of samples of the process, e.g., fix a number of samples n and a collection of sample times $t_0, t_1, \ldots, t_{n-1}$. We can use the formula for finding the distribution of a random vector to find the distribution for this collection of random variables, e.g., considering cdf's we have that

$$F_{X_{t_0},\ldots,X_{t_{n-1}}}(\mathbf{x}) = P(X_{t_i} \leq x_i;\ i = 0, \ldots, n-1).$$

Furthermore, all of these cdf's, that is, the cdf's for all choices of numbers of samples and sample times, must be *consistent* in the sense that we cannot get different values for the probability of a fixed event using different cdf's. For example, we could find $\Pr(X_0 \leq 3)$ as $F_{X_0}(3)$ or $F_{X_0,X_7}(3, \infty)$ and all answers must agree. The Kolmogorov extension theorem is essentially a converse to this result. It states that if we have a consistent family of distributions (or, equivalently, of cdf's or pdf's or pmf's) for all possible collections of finite samples of the process, then the process is uniquely specified. In other words, given a consistent description of finite dimensional distributions, then there exists one and only one random process which possesses those finite dimensional distributions. This provides a powerful tool for constructing models of random processes. Two of the most important examples follow. A discrete time process is said to be *independent and identically distributed* or *iid* if there is a cdf F such that all n-dimensional cdf's of the process are product cdf's with marginal cdf F. The name follows since the condition implies that the random variables produced by the process are independent and have identical distributions. Thus all of the pdf's and pmf's thus far considered yield corresponding iid processes. It is straightforward to show that the resulting distributions are consistent and hence the processes well defined.

A process is said to be *Gaussian* if all finite collections of samples are Gaussian random vectors. It is straightforward (but tedious) to show that the consistency conditions are met if the following conditions are met:

1. There exists a function $\{m(t); t \in \mathcal{T}\}$, called the *mean function*, and a function $\{\Lambda(t,s);\ t,s \in \mathcal{T}\}$, called the *covariance function*, which is symmetric and strictly positive definite, i.e., $\Lambda(t,s) = \Lambda(s,t)$ and for any n and $t_1, \ldots, t_n$, and real $(y_0, \ldots, y_{n-1})$ we have that

$$\sum_{i=0}^{n-1}\sum_{j=0}^{n-1} y_i \Lambda(t_i, t_j) y_j > 0.$$

2. Given n and $t_0, \cdots, t_{n-1}$ the vector $(X_{t_0}, \ldots, X_{t_{n-1}})$ is Gaussian with $(m_{t_0}, \ldots, m_{t_{n-1}})$ and covariance $\{\Lambda(t_i, t_j);\ i,j = 0, \ldots, n-1\}$.

A discrete time Gaussian process is iid if and only $m(t)$ is a constant, say m, and $\Lambda(t,s)$ has the form $\sigma^2 \delta_{t-s}$, where δ_k is the Kronecker delta, 1 if $k = 0$ and 0 otherwise.

It is often useful to consider conditional probabilities when specifying a random process. A few remarks will provide sufficient machinery for our purposes. Suppose that we fix n and a collection of sample times $t_0 < t_1 < \ldots < t_{n-1}$. Suppose that the process is described by pmf's and consider the pmf for the random vector $X_{t_0}, X_{t_1}, \ldots, X_{t_{n-1}}$. We can write

$$p_{X_{t_0},X_{t_1},\ldots,X_{t_{n-1}}}(x_0, x_1, \ldots, x_{n-1}) =$$
$$p_{X_{t_0}}(x_0) \prod_{k=1}^{n-1} \frac{p_{X_{t_0},X_{t_1},\ldots,X_{t_k}}(x_0, x_1, \ldots, x_k)}{p_{X_{t_0},X_{t_1},\ldots,X_{t_{k-1}}}(x_0, x_1, \ldots, x_{k-1})}.$$

The ratios in the above product are recognizable as elementary conditional probabilities, and hence we define the conditional pmf's

$$p_{X_{t_k}|X_{t_{k-1}},\ldots,X_{t_0}}(x_k|x_{k-1}, \ldots, x_0) = \Pr(X_{t_k} = x_k | X_{t_i} = x_i; i = 0, \ldots, k-1)$$
$$= \frac{p_{X_{t_0},X_{t_1},\ldots,X_{t_k}}(x_0, x_1, \ldots, x_k)}{p_{X_{t_0},X_{t_1},\ldots,X_{t_{k-1}}}(x_0, x_1, \ldots, x_{k-1})}.$$

This can be viewed as a generalization of a product measure and a chain rule for pmf's. By analogy we can write a similar form in the continuous case. Suppose now that the process is described by pdf's. We can write

$$f_{X_{t_0},X_{t_1},\ldots,X_{t_{n-1}}}(x_0,x_1,\ldots,x_{n-1}) = \\ f_{X_{t_0}}(x_0)\prod_{k=1}^{n-1}\frac{f_{X_{t_0},X_{t_1},\ldots,X_{t_k}}(x_0,x_1,\ldots,x_k)}{f_{X_{t_0},X_{t_1},\ldots,X_{t_{k-1}}}(x_0,x_1,\ldots,x_{k-1})}.$$

Define the conditional pdf's by

$$f_{X_{t_k}|X_{t_{k-1}},\ldots,X_{t_0}}(x_k|x_{k-1},\ldots,x_0) = \frac{f_{X_{t_0},X_{t_1},\ldots,X_{t_k}}(x_0,x_1,\ldots,x_k)}{f_{X_{t_0},X_{t_1},\ldots,X_{t_{k-1}}}(x_0,x_1,\ldots,x_{k-1})}.$$

These are no longer probabilities (they are *densities* of probability), but we can relate them to probabilities as follows: Define the conditional probability

$$P(X_{t_k}\in G|X_{t_i}=x_i; i=0,\ldots,k-1) = \\ \int_G f_{X_{t_k}|X_{t_{k-1}},\ldots,X_{t_0}}(x_k|x_{k-1},\ldots,x_0)\,dx_k.$$

Thus conditional pdf's can be used to compute conditional probabilities.

A simple example of the use of conditional probabilities to describe a process is in the definition of Markov processes. A *Markov process* is one for which for any n and any $t_0 < t_1 < \cdots < t_{n-1}$ we have for all $x_0,\ldots,x_{n-1}$ that $\Pr(X_{t_n}\in G_n|X_{t_{n-1}}=x_{n-1},\ldots,X_{t_0}=x_0) = \Pr(X_{t_n}\in G_n|X_{t_{n-1}} = x_{n-1})$; that is, given the recent past, probabilities of future events do not depend on previous outcomes. If the process is described by pdf's this is equivalent to

$$f_{X_{t_n}|X_{t_{n-1}},\ldots,X_{t_0}}(x_n|x_{n-1},\ldots,x_0) = f_{X_{t_n}|X_{t_{n-1}}}(x_n|x_{n-1}).$$

If the process is described by pmf's, it is equivalent to

$$p_{X_{t_n}|X_{t_{n-1}},\ldots,X_{t_0}}(x_n|x_{n-1},\ldots,x_0) = p_{X_{t_n}|X_{t_{n-1}}}(x_n|x_{n-1}).$$

If a process is Gaussian and has a covariance of the form $\Lambda(t,s)=\sigma^2\rho^{-|t-s|}$, then it is Markov. This process is usually called the *Gauss Markov process*. If the process is Gaussian and defined only for nonnegative time and has a covariance of the form $\Lambda(t,s)=\sigma^2\min(t,s)$, then it is also Markov. This process is called the Wiener process.

1.4 Expectation

Suppose that X is a random variable described by either a pdf f_X or pmf p_X. The *expectation* of X is defined by the integral

$$EX = \int x f_X(x)\, dx$$

if described by a pdf and by the sum

$$EX = \sum_x x p_X(x)$$

if described by a pmf. Given a random variable X defined on a probability space $(\Omega, \mathcal{F}, P)$, we can define another random variable, say g, which is a function of X. That is, the composition of the two random variables X and g yields a new random variable $Y = g(X)$ defined on the original space by $Y(\omega) = g(X)(\omega) = g(X(\omega))$. In other words, we form the measurement Y on Ω by first taking the measurement $X(\omega)$ and then taking g on the result. We could find the expectation of Y by first finding its cdf F_Y and then taking the appropriate integral with respect to the cdf, but the *fundamental theorem of expectation* provides a shortcut and allows us to use the original pdf or pmf. For example, if the appropriate pdf's are defined,

$$EY = Eg(X) = \int y f_Y(y)\, dy = \int g(x) f_X(x)\, dx.$$

Expectations of a random variable with respect to an event are frequently used. If $A \in \Omega$ is an event, then

$$E(X|A) = \int x f_{X|A}(x|A)\, dx.$$

Let $\{A_i\}$ be a sequence of events that form a partition of the sample space so that

$$\bigcup_i A_i = \Omega \ , \ A_i \bigcap A_j = \emptyset \text{ for } i \neq j,$$

then

$$EX = \sum_i E(X|A_i) \Pr(A_i).$$

Expectations can also be taken with respect to conditional probability functions to form conditional expectations, e.g., given two continuous random variables X and Y

$$E(Y|x) = E(Y|X = x) = \int y f_{Y|X}(y|x)\, dy.$$

It should be clearly understood that $E(Y|x)$ is a deterministic function of x. Then, by replacing x by the random variable X, the expression $E(Y|X)$ defines a new random variable that is a function of the random variable X. Conditional expectations are useful for computing ordinary expectations using the so-called iterated expectation or nested expectation formula:

$$E(Y) = \int y f_Y(y)\, dy = \int y \left(\int f_{X,Y}(x,y)\, dx \right) dy$$

$$= \int f_X(x) \left(\int y f_{Y|X}(y|x)\, dy \right) dx = \int f_X(x) E(Y|x)\, dx = E(E(Y|X)).$$

Two important examples of expectation of functions of random variables are the *mean*

$$m_X = EX$$

and the *variance*

$$\sigma_X^2 = E[(X - m_X)^2].$$

Expectations of the form $E(X^n)$ are called the nth order *moments* of the random variable. Note that it is usually easier to evaluate the variance and moments using the fundamental theorem of expectation than to derive the new cdf's. Suppose that $\{X_t;\ t \in \mathcal{T}\}$ is a random process. The *mean function* or *mean* is defined by

$$m_X(t) = EX_t.$$

The *autocorrelation function* or *autocorrelation* of the process is defined by

$$R_X(t,s) = E(X_t X_s).$$

The *autocovariance function* or *covariance* is defined by

$$K_X(t,s) = E((X_t - m_X(t))(X_s - m_X(s))).$$

Observe that

$$K_X(t,t) = \sigma_{X_t}{}^2.$$

It is easy to see that

$$K_X(t,s) = R_X(t,s) - m_X(t) m_X(s).$$

A process $\{X_t; t \in \mathcal{T}\}$ is said to be *weakly stationary* if

1. $m_X(t) = m_X(t+\tau)$, all t, τ such that $t, t+\tau \in \mathcal{T}$, and

2. $R_X(t,s) = R_X(t+\tau, s+\tau)$, all τ such that $t, s, t+\tau, s+\tau \in \mathcal{T}$.

A process is weakly stationary if the first two moments—the mean and the correlation—are unchanged by shifting, that is, by changing the time origin. This is equivalent to saying that the mean function is constant, that is, there is a constant m_X such that

$$m_X(t) = m_X,$$

(does not depend on time); and that the correlation depends on the two sample times only through their difference, a fact that is often expressed as

$$R_X(t,s) = R_X(t-s).$$

A function of two variables which depends on the variables only through their difference as above is said to be *Toeplitz*. Weak stationarity is a form of *stationarity property* of a random process. Stationarity properties refer to moments or distributions of samples of the process being unchanged by time shifts. We shall later consider other stationarity properties. If a process $\{X_t;\, t \in \mathcal{T}\}$ is weakly stationary, then we define its *power spectral density* $S_X(f)$ as the Fourier transform of the correlation:

$$S_X(f) = \mathcal{F}(R_X(t)) = \begin{cases} \int R_X(t) e^{-j2\pi f t}\, dt & \text{if } \mathcal{T} \text{ continuous} \\ \sum R_X(k) e^{-j2\pi f k} & \text{if } \mathcal{T} \text{ discrete} \end{cases}$$

1.5 Ergodic Properties

We have already seen an example of a stationarity property of a process: A process $\{X_t\}$ is weakly stationary if the mean function and autocorrelation function are not affected by time shifts, that is,

$$EX_t = EX_{t+\tau}$$

and

$$R_X(t,s) = R_X(t+\tau, s+\tau),$$

for all values of t, s, τ for which the above are defined. While the same thing can be stated more concisely as $EX_t = m$ and $R_X(t,s) = R_X(t-s)$, the above form better captures the general idea of stationarity properties. We say that a process is *strictly stationary* or, simply, *stationary* if the probabilities of all events are unchanged by shifting the events. A precise mathematical definition is complicated, but the condition is equivalent to the following: For any dimension n and any choice of n sample times

$t_0, \ldots, t_{n-1}$, the joint cdf for the n samples of the process at these times satisfies

$$F_{X_{t_0},\ldots,X_{t_{n-1}}}(x_0, \ldots, x_{n-1}) = F_{X_{t_0+\tau},\ldots,X_{t_{n-1}+\tau}}(x_0, \ldots, x_{n-1})$$

for all choices of τ and $\mathbf{t}$ for which the cdf's are defined. Strict stationarity is also equivalent to the formulas obtained by replacing the cdf's above by pdf's or pmf's, if appropriate. The key idea is that the probabilistic description of the random process is unchanged by time shifts. There are more general stationarity properties than the strict stationarity defined above. For example, processes can have nonstationarities due to transients that die out in time or periodicities. In the first case the distributions of samples converge to a stationary distribution as the sample times increase and in the second case a stationary distribution results if the distributions are suitably averaged over the period. Making these ideas precise lead to the notions of *asymptotically stationary* and *asymptotically mean stationary*, which we mention solely to point out that there is nothing "magic" about the property of being stationary: many of the useful properties of stationary processes hold also for the more general notions.

There are two uses of the word *ergodic* in describing a random process. The first, that of an *ergodic process*, is a rather abstract mathematical definition, while the second, that of the *ergodic property*, describes a key property of a random process that is important for both theory and application. The first concept is quite technical and would require additional notation and terminology to define precisely. Roughly speaking it states that the process has the property that if a set of signals is unaffected by shifts in time, that is, shifting a signal in the set gives another signal in the set, then that set must have either probability 1 or probability 0. If the process is also stationary (or asymptotically stationary), it can be shown that this technical condition is equivalent to the property that events that are widely separated in time are approximately independent. In other words, ergodicity is a condition on the decay of the memory of a random process.

The concept of an *ergodic property* is more operational. As already discussed, a fundamental goal of probability theory is to relate the long term behavior of random processes to computable quantities, e.g., to relate sample or time averages to expectations. Suppose for example that $\{X_n\}$ is a real-valued discrete time random process. It is of interest to know conditions under which the sample mean

$$\frac{1}{n}\sum_{k=0}^{n-1} X_k$$

converges in some sense as $n \to \infty$. Other limits may also be of interest, such as the sample autocorrelation

$$\frac{1}{n}\sum_{k=0}^{n-1} X_k X_{k+n}$$

or the sample average power

$$\frac{1}{n}\sum_{k=0}^{n-1} X_k^2.$$

We might pass $\{X_n\}$ through a linear filter with pulse response h_k to form a new process $\{Y_n\}$ and then ask if

$$\frac{1}{n}\sum_{k=0}^{n-1} Y_k$$

converges. In general, we could take a series of measurement g_n on the random process (such as the Y_n above) which could in principle depend on the entire past and future and ask if the sample average

$$\frac{1}{n}\sum_{k=0}^{n-1} g_k$$

converges. If the process has continuous time, then we are interested in the time averages obtained by replacing the above sums by integrals, e.g., the limit as $T \to \infty$ of

$$\frac{1}{T}\int_0^T X_t dt,$$

and so on. A random process is said to have the *ergodic property* if *all* such time averages for reasonable measurements on the process converge in some sense. The type of convergence determines the type of ergodic property, e.g., convergence in mean square, in probability, or with probability one. Theorems proving that a given process has the ergodic property are called *ergodic theorems* or, occasionally, *laws of large numbers.* Perhaps the most famous ergodic theorem states that if a discrete time random process is stationary, then it has the ergodic property. If the process is also ergodic, then with probability one sample averages converge to expectations, e.g.,

$$\lim_{n\to\infty} \frac{1}{n}\sum_{k=0}^{n-1} X_k = E(X_0).$$

If the process is not ergodic, then sample averages still converge, but they will converge in general to a random variable. It is important to observe that stationarity is sufficient for the convergence of sample averages, but it is not necessary. Similarly, ergodicity alone is not sufficient for the convergence of sample averages, but combined with stationarity it is. It is erroneous to assume that either stationarity or ergodicity are required for a process to possess the ergodic property, that is, to have convergent time averages. For example, processes can have transients or short term behavior which will cause them to be nonstationary, yet sample averages will still converge.

Processes may have distinct ergodic modes and hence not be ergodic, e.g., the process may be Gaussian with a known correlation but with a time-independent mean function chosen at random from [0,1] according to an unknown distribution. This process can be viewed as a mixture of ergodic processes, but it is not itself ergodic. It does, however, have the ergodic property and time averages will converge. In fact all stationary processes that are not ergodic can be modeled as a mixture of ergodic sources, that is, as a doubly stochastic process where nature first selects a process at random from a collection of ergodic processes, and then produces the output of that process forever. If we observe a stationary nonergodic process, then we are actually observing an ergodic process, we just don't know which one. We can determine which one, however, by observing the long term behavior and estimating the probabilities from relative frequencies. This property of stationary processes is called the *ergodic decomposition* and it eases the application of random process tools to nonergodic processes. As a final comment on the ergodic property, we note that it can be shown that a sufficient *and necessary* condition for a random process to have the ergodic property is that it be asymptotically mean stationary, the generalization of stationary mentioned previously. Happily virtually all information sources and processes produced by all coding structures encountered in the real world are well modeled by asymptotically mean stationary processes. Furthermore, for any asymptotically mean stationary source there is an equivalent stationary source with the same long term sample averages. For this reason it is often reasonable to assume that processes are stationary in analysis even if they are not, provided that they do satisfy the more general conditions.

Quasi-stationary Processes

We will occasionally need to consider spectral analysis for non-stationary processes. In this section we review a class of processes which can be considered as the most general class of processes for which spectral analysis makes sense. This class, called *quasi-stationary* processes, provides a nice generalization of both well-behaved random processes and important deterministic processes. A detailed treatment may be found in Ljung [60]. A discrete time process X_n is said to be *quasi-stationary* if there is a finite constant C such that

$$E(X_n) \leq C; \text{ all } n$$

$$|R_X(n,k)| \leq C; \text{ all } n, k$$

where $R_X(n,k) = E(X_n X_k)$, and if for each k the limit

$$\lim_{N\to\infty} \frac{1}{N} \sum_{n=1}^{N} R_X(n, n+k) \tag{1.5.1}$$

exists, in which case the limit is defined as $R_X(k)$. For technical reasons we add the first moment condition that the limit

$$\bar{m}_X = \lim_{N\to\infty} \frac{1}{N} \sum_{n=1}^{N} E(X_n)$$

exists to avoid the implicit assumption of a zero mean. Following Ljung we introduce the following notation: Given a process X_n, define

$$\bar{E}\{X_n\} = \lim_{N\to\infty} \frac{1}{N} \sum_{n=1}^{N} E(X_n), \tag{1.5.2}$$

if the limit exists. Thus for a quasi-stationary process $\{X_n\}$ the autocorrelation is given by

$$R_X(k) = \bar{E}\{ X_n X_{n+k}\}, \tag{1.5.3}$$

the mean is defined by

$$m_X = \bar{E}\{X_n\} \tag{1.5.4}$$

and the average power is given by

$$R_X(0) = \bar{E}\{X_n^2\}. \tag{1.5.5}$$

These moments reduce to the corresponding time averages or probabilistic averages in the special cases of deterministic or random processes, respectively.

The fundamental common examples of quasi-stationary processes are the class of deterministic processes with convergent and bounded sample means and autocorrelations and the class of stationary random processes (with bounded mean and autocorrelation). The autocorrelation defined by

1.5.3 reduces to the sample average in the former case and the usual probabilistic autocorrelation in the latter. Useful but still simple examples are formed by combining simple deterministic processes with stationary random processes, e.g., adding a sinusoid to an iid process. More general examples of quasi-stationary random processes include processes that are second order stationary processes (second order distributions do not depend on time origin) or asymptotically mean stationary processes (processes for which the ergodic theorem holds for all bounded measurements on the process) provided that the appropriate moments are bounded.

The *power spectrum* of the process is defined in the general case as the discrete time Fourier transform of the autocorrelation:

$$S_X(f) = \sum_{n=-\infty}^{\infty} R_X(n) e^{-j2\pi fn}, \tag{1.5.6}$$

where the frequency f is normalized to lie in $[-1/2, 1/2]$ or $[0,1]$ (corresponding to a sampling period being considered as a single time unit). The usual linear system input/output relations hold for this general definition of spectrum (see Chapter 2 of Ljung [60]).

Exercises

1. Given a random variable X, two random variables W and V (considered to be noise) are added to X to form the observations

$$Y = X + W$$

$$Z = X + V.$$

   Both W and V are assumed to be independent of X. Suppose that all variables have 0 mean and that the variance of X is σ_X^2 and the random variables W and V have equal variances, say σ^2. Lastly, suppose that

$$E(WV) = \rho\sigma^2.$$

   (a) What is the largest possible value of $|\rho|$? (Hint: Look up the Cauchy-Schwartz inequality.)

   (b) Find $E(YZ)$.

   (c) Two possible estimates for the underlying random variable X based on the observations are

$$\hat{X}_1 = Y$$

and

$$\hat{X}_2 = \frac{Y + Z}{2},$$

that is, one estimate looks at only one observation and the other looks at both. For what values of ρ is $\hat{X}_2$ the better estimate in the sense of minimizing the mean squared error

$$E[(\hat{X}_i - X)^2]?$$

For what value of ρ is this best mean squared error minimized? What is the resulting minimum value?

(d) Suppose now that X, W, V are jointly Gaussian and that ρ is 0. (We still assume that W, V are independent of X.) Find the joint pdf

$$f_{Y,Z}(y, z).$$

(e) Suppose that ρ is arbitrary, but the the random variables are Gaussian as in part (d). Are the random variables $Y + Z$ and $Y - Z$ independent?

2. Let $\{Y_n\}$ be a two-sided stationary process defined by the difference equation

$$Y_n = \rho Y_{n-1} + X_n,$$

where $\{X_n\}$ is an iid process, $|\rho| < 1$, and the X_n are independent of previous Y_k; $k < n$. Suppose the we know that

$$f_{Y_n}(y) = \frac{\lambda}{2} e^{-\lambda |y|}, \text{ all } y,$$

a doubly exponential pdf. (This is a popular model for sampled speech, where ρ is typically between .85 and .95, depending on the sampling rate.)

(a) Find $E(X)$, σ_X^2, and the autocorrelation $R_X(k, j)$.

(b) Find the *characteristic function* $\Phi_X(ju) = E(e^{jux})$.

(c) Find the probability $\Pr(Y_n X_n + 1/2)$ and the conditional pdf $f_{X_n|Y_n}(x|y) = f_{X_n,Y_n}(x, y)/f_{Y_n}(y)$.

(d) What is the power spectral density of $\{Y_n\}$?

(e) Find $f_{Y_n|Y_{n-1}}(y|z)$ and $E(Y_n|Y_{n-1})$.

3. A continuous time Gaussian random process $\{X(t);\ t \in (-\infty, \infty)\}$ is described by a mean function $E[X(t)] = 0$ all t and an autocorrelation function $R_X(t,s) = E(X(t)X(s)) = \sigma^2 \rho^{-|t-s|}$ for a parameter ρ in $[-1, 1]$. The process is digitized in two steps as follows: First the process is sampled every T seconds to form a new discrete time random process $X_n = X(nT)$. The samples X_n are then put through a binary quantizer defined by

$$q(x) = \begin{cases} +a & \text{if } x \geq 0 \\ -a & \text{if } x < 0 \end{cases}$$

to form a new process $q(X_n)$.

   (a) Write an expression for the pdf

   $$f_{X_n, X_k}(x, y)$$

   (b) Find an expression for the mean-squared quantization error defined by

   $$\Delta = E[(X_n - q(X_n))^2]$$

   What choice of a minimizes Δ? (Hint: Use calculus.)

   (c) Suppose now that $\rho = 1$ above which means that

   $$E(X_n X_k) = E(X_0^2)$$

   for all n and k. Is there a random variable Y such that

   $$\frac{1}{n} \sum_{i=0}^{n-1} X_i$$

   converges in mean square to Y, i.e.,

   $$\lim_{n \to \infty} E[(\frac{1}{n} \sum_{i=0}^{n-1} X_i - Y)^2] = 0?$$

4. Let $\{X_n\}$ be an iid sequence of random variables with a uniform probability density function on $[0, 1]$. Suppose we define an N-level uniform quantizer $q(r)$ on [0,1] as follows: $q(r) = (k + 1/2)/N$ if $k/N \leq r < (k + 1)/N$. In other words, we divide [0,1] up into N intervals of equal size. If the input falls in a particular bin, the output is the midpoint of the bin. Application of the quantizer to the process $\{X_n\}$ yields the process $\{q(X_n)\}$. Define the quantizer error process $\{\epsilon_n\}$ by $\epsilon_n = q(X_n) - X_n$.

(a) Find $E(\epsilon_n)$, $E(\epsilon_n^2)$, and $E(\epsilon_n q(X_n))$.

(b) Find the (marginal) cumulative distribution function and the pdf for ϵ_n.

(c) Find the covariance and power spectral density of ϵ_n. *Hint:* Break up the integrals into a sum of integrals over the separate bins.

5. A k-dimensional random vector $\mathbf{X}$ is coded as follows: First it is multiplied by a unitary matrix $\mathbf{U}$ to form a new vector $\mathbf{Y} = \mathbf{UX}$. (By unitary it is meant that $\mathbf{U}^* = \mathbf{U}^{-1}$, where $\mathbf{U}^*$ is the complex conjugate of $\mathbf{U}$). Each component Y_i; $i = 0, 1, \ldots, k-1$ is separately quantized by a quantizer q_i to form a reproduction $\hat{Y}_i = q_i(Y_i)$. Let $\hat{\mathbf{Y}}$ denote the resulting vector. This vector is then used to produce a reproduction $\hat{\mathbf{X}}$ of $\mathbf{X}$ by the formula $\hat{\mathbf{X}} = \mathbf{U}^{-1}\hat{\mathbf{Y}}$. This code is called a *transform code*.

(a) Suppose that we measure the distortion between input and output by the mean squared error

$$(\mathbf{X} - \hat{\mathbf{X}})^t(\mathbf{X} - \hat{\mathbf{X}}) = \sum_{i=0}^{k-1} |X_i - \hat{X}_i|^2.$$

Show that the distortion is the same in the original and transform domain, that is,

$$(\mathbf{X} - \hat{\mathbf{X}})^t(\mathbf{X} - \hat{\mathbf{X}}) = (\mathbf{Y} - \hat{\mathbf{Y}})^*(\mathbf{Y} - \hat{\mathbf{Y}}).$$

(b) Find a matrix $\mathbf{U}$ such that $\mathbf{Y}$ has uncorrelated components, that is, its covariance matrix is diagonal.

6. Consider the coding scheme of Figure 1.1 in which the encoder is a simple quantizer inside a feedback loop. H and G are causal linear filters. Show that if the filters are chosen so that

$$\hat{X}_n = \tilde{X}_n + q(\epsilon_n),$$

then

$$|X_n - \hat{X}_n| = |\epsilon_n - q(\epsilon_n)|$$

and hence

$$E(|X_n - \hat{X}_n|^2) = E(|\epsilon_n - q(\epsilon_n)|^2),$$

that is, the overall error is the same as the quantizer error. Suppose that G is fixed. What must H be in order to satisfy the given condition?

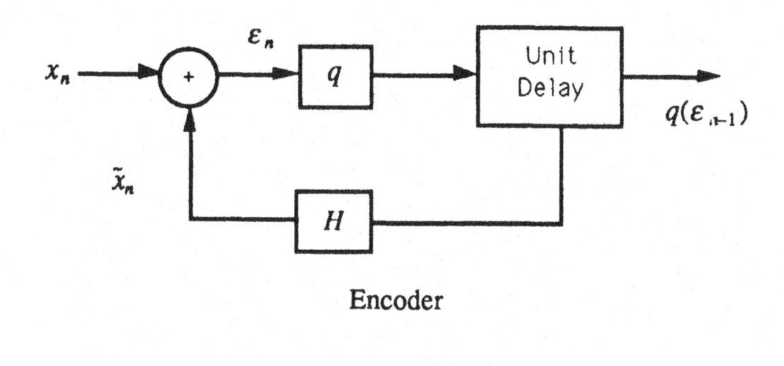

Encoder

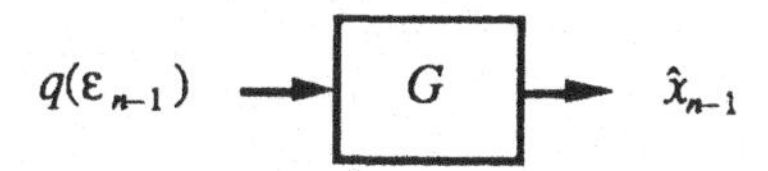

Decoder

Figure 1.1: Feedback Quantizer

7. Find the most reputable uniform number generator on your computer. Test the uniformity of the random number generator by finding the relative frequencies for a sequence of 10,000 outputs lying in two equally sized subintervals of the overall output range (e.g., if the uniform generator produces numbers in $[0, 1)$, find the histograms for the intervals $[0, 1/2)$ and $[1/2, 1)$. Repeat for 4 equally sized intevals and 8 equally sized intervals. Find the sample autocorrelation for lags 0 through 4 to see how "white" the random number generator is.

8. Suppose that you have a perfect uniformly distributed random number generator that provides a samples of a random variable U having a uniform distribution on $[0, 1)$. How can you use this to generate samples of a random variable X having some specified distribution $F(x)$? *Hint:* Consider the inverse distribution $F^{-1}(r)$; $r \in [0, 1]$ defined as the smallest value of $x \in \mathcal{R}$ for which $F(x) = r$. Define a new random variable $Z = F^{-1}(U)$ and derive its distribution.

Chapter 2

Codes, Distortion, and Information

2.1 Basic Models of Communication Systems

Information theory, the mathematical theory of communication, originated with the classical papers of Claude Shannon [73], [74]. Shannon modeled communication systems as random processes, introduced probabilistic definitions of *information*, and used classical ergodic theorems together with new ergodic theorems for information to develop coding theorems, results which quantified the optimal achievable performance in idealized communication systems. Shannon considered two fundamental aspects of communication: *source coding* or the mapping of an information source into a representation suitable for transmission on a channel and *channel coding* or error control coding for the reliable communication over a noisy channel. Our emphasis has been on and will remain on source coding.

There is no single mathematical model of a communication system that captures every aspect of the corresponding physical system. The more one tries to include, the more complicated the model becomes. Hence we aim for a model that is sufficiently rich to capture the most important aspects of a real system, yet is sufficiently simple to permit useful analysis and theory. Many aspects of significance to some applications will be buried in the models considered here, receiving at best only passing mention.

We begin with a model shown in Fig. 2.1 that is quite general and is the starting point for most books on communication and information theory. A communication system consists of a source, which is just a discrete time random process $\{X_n\}$, an encoder, a channel, a decoder, and a destination

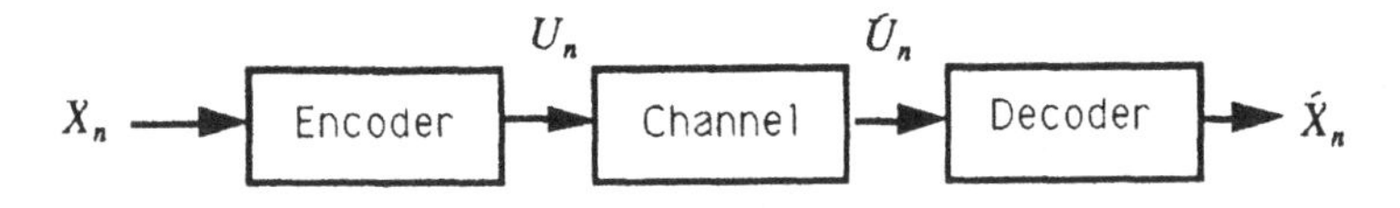

Figure 2.1: Communication System

or received random process $\{\hat{X}_n\}$, which should be an approximation in some sense to the input source $\{X_n\}$. In general we permit the source to be a vector-valued process, e.g., X_n may take values in k-dimensional Euclidean space, $\mathcal{R}^k$. We will not usually use boldface to denote vectors, although it will occasionally be done for emphasis. We shall usually let A denote the sample space of possible values of X_n and $\hat{A}$ be the sample space of possible values of the reproduction $\hat{X}_n$. These two spaces are often considered to be the same, but not always.

If the source produces one symbol every τ_s seconds, then we shall refer to $R_{source} = 1/\tau_s$ as the *source symbol rate* in *symbols per second*. If the discrete time source is obtained from a continuous time source by sampling, then the source symbol rate is just the sampling rate and we speak of samples instead of symbols. The channel is usually considered to be that part of the system which is beyond the control of the designer. It can include deterministic operations such as filtering, modulation, Doppler shifts, multipath, and aliasing. It can include random effects such as fading, additive noise, and jamming. In most treatments of source coding, the channel is by and large ignored. This is justified because of Shannon's joint source and channel block coding theorem which implies that one can consider the two coding operations separately. Although we do not consider error control or channel coding here, we do devote some time to the constraints that a channel places on a source code. In particular, we briefly relate several notions of rate (code rate and transmission rate) and discuss the necessity for matching the output rate of the source with the input rate of the channel. While this discussion can be avoided by invoking the separation of source and channel codes, it is included because of the insight it sheds on connecting source coding systems to communication channels and the bounds imposed on source coding by the physical limits of communication or storage channels.

Our basic assumptions on the channel are twofold: First, it is digital; its permissible inputs and outputs are in a finite collection of possible symbols. We often assume that the channel alphabet is a space of binary vectors. Second, it has a fixed rate, that is, a symbol comes out of the channel for every symbol that goes in. Symbols cannot be lost or added by the channel. Thus for every channel input symbol U_n, exactly one channel output symbol $\hat{U}_n$

is produced. As in any compromise between mathematics and reality, this assumption omits some interesting special cases where insertions and deletions are as important as errors. In general, a channel is described by a family of conditional probability measures, e.g., the measures for output sequences given input sequences. When studying data compression systems, however, it is usually assumed that the channel is *noiseless* in the sense that the output symbol is the same as the input symbol, $\hat{U}_n = U_n$ for all n. The theoretical justification for this assumption is the *data transmission theorem* which states that one can design separate systems for data compression (source coding) and error control (channel coding) and then simply combine these systems in cascade to obtain a good overall system. In other words, using error correction coding one can replace a noisy channel by a noiseless channel of a smaller rate. A practical justification for the assumption of a noiseless channel is that it permits the designer to focus on the data compression problem alone. We usually make the noiseless channel assumption, but we emphasize that channel errors can have disastrous effects on data compression systems and the possibility must be considered in any real system design, if only by simulating the system to quantify the effects of channel errors. We will comment occasionally on the effects of such errors on different types of data compression systems.

Suppose that the channel alphabet is B and there are $||B||$ symbols in the alphabet. If the channel accepts and transmits one symbol in B each τ_c seconds and hence $L = 1/\tau_c$ symbols each second, then we shall define the *channel rate* $R_{channel}$ as L symbols per second. It is usually more convenient to normalize such rates and hence we also consider R_{channel} to be $L \log_2 ||B||$ *bits per second* or $L \ln ||B||$ *nats per second*. The first definition is a form of normalization since it states that regardless of the actual channel alphabet, R_{channel} binary symbols must be sent per second over a period of time in order to characterize the channel inputs during that time period. The second definition is also a form of normalization, but it is more useful for developing the theory since some tools are most easily expressed in terms of natural logarithms.

The encoder can be viewed as a device for translating the original source $\{X_n\}$ into a sequence $\{U_n\}$ acceptable by the channel, that is, a sequence of symbols drawn from the finite alphabet of the channel. The decoder converts the channel symbol sequence into a form appropriate for approximating the original input source. In general, the encoder operates on a sequence of input symbols and produces a sequence of channel symbols. There are, however, a variety of methods by which such a mapping can be accomplished and we shall see many possible encoder structures. The notion of time for the input source may not be the same as that of the channel, that is, it may or may not be true that one channel input symbol is produced by the encoder for each source symbol input to the encoder. We do require, however, that the encoder produce an output sequence of channel symbols at the channel rate. In a similar manner, the decoder must map a sequence of channel symbols into a sequence of reproduction symbols. The emphasis of this book will be on *fixed rate* systems, systems wherein the coders accept input symbols and produce output symbols at a fixed rate. For such systems we assume that there are integers k and K such that the time required for the encoder to produce K channel symbols is the same as that required by the source to produce k symbols. This is in contrast to variable rate or variable length codes where, for example, input words of fixed length can produce output words of different lengths. While we shall see examples of such systems, the bulk of the information theoretic development will assume fixed rate systems. In the case of a fixed rate system producing K channel symbols at a rate of R_{channel} symbols per second in the same time it takes the source to produce k source symbols at a rate of R_{source} symbols per second, we must have that

$$K R_{\text{source}} = k R_{\text{channel}}.$$

Equivalently, the result follows since 1 second equals both $k\tau_s = k/R_{\text{source}}$ and $K\tau_c = K/R_{\text{channel}}$. Hence we can define the quantity *transmission rate* of the system as the number of channel bits (or nats) available for each source symbol, that is,

$$R_{\text{trans}} = \frac{R_{\text{channel}}}{R_{\text{source}}} = \frac{K}{k} \log ||B|| = \frac{\tau_s}{\tau_c} \log ||B||. \tag{2.1.1}$$

This quantity does not explicitly involve time (which is implicitly contained in the fact that k source symbols and K channel symbols occupy the same amount of time). While a similar quantity can be defined for some variable rate systems by considering the average number of channel bits per source sample, the definition is more complicated to use and does not always have the desired intuition.

2.2 Code Structures

In this section we describe several code structures. The simplest structure is a *fixed-length block code* or *vector code*: Here we parse the input sequence into nonoverlapping groups or vectors of dimension, say, k. Each successive input k dimensional vector is mapped into a vector of channel symbols of dimension, say, K. In more detail, each input vector

$$\mathbf{X}_n^t = (X_{nk}, X_{nk+1}, \ldots, X_{(n+1)k-1})$$

is mapped into a channel vector

$$\mathbf{U}_n^t = (U_{nK}, U_{nK+1}, \ldots, U_{(n+1)K-1})$$

for all integers n. The superscript t denotes transpose and is used since $\mathbf{X}$ is considered to be a column vector. If the channel is noiseless, this can be called a *vector quantizer* (or VQ) because it maps vectors from a possibly continuous space into a finite set. The name vector quantizer is more properly applied to the special case where the encoder is a minimum distortion or nearest neighbor mapping, as will be seen later.

If a block code decoder is used to map channel K-blocks into reproduction k-blocks, then the overall system maps successive k-blocks from the source into k-dimensional reproduction vectors. In this case it is straightforward to show that if the input process is stationary, then the resulting input/output process is k-stationary. The same conclusion holds if the input process is only k-stationary. Such vector codes are said to be *memoryless* because each input vector is mapped into a code vector without regard to past or future input vectors or output codewords.

In the special case where the input dimension is k=1, the code is called a *scalar code*. The classic example of a scalar code is a *scalar quantizer*, a vector quantizer of dimension 1. Suppose that the $\{X_n\}$ are real numbers and that we have a sequence of thresholds $-\infty \leq b_0 < b_1 < \cdots \leq b_M < \infty$ and a sequence of reproduction values $-\infty < a_0 < a_1 < \cdots < a_{M-1} < \infty$ Then a sequence of source symbols $\{X_n\}$ can be encoded into a sequence of symbols from $\{0, 1, \ldots, M-1\}$ by an encoder α defined by $\alpha(X_n) = i$ if $b_i < X_n \leq b_{i+1}$ and a sequence of symbols in $\{0, 1, \ldots, M-1\}$ can be decoded by a decoder β defined by the mapping $\beta(U_n) = a_i$ if $U_n = i$; that is, the received channel symbol can be viewed as an index for the reproduction symbol stored in memory and hence the decoder is a scalar table lookup. The overall mapping can be depicted as in Fig. 2.2, which shows a quantizer input/output mapping for the special case of 8 equally separated output levels, a so-called uniform quantizer.

Note that we could equally well model the channel as consisting of scalars drawn from $\{0, 1, \ldots, M-1\}$ or as $\log_2 M$-dimensional binary vectors. We refer to the code as a *scalar* code because the input symbols are treated as scalars and not grouped as vectors.

One generalization of the memoryless vector code described above is to permit the dimension of the encoded word to vary, that is, the length of a codeword can depend on the input word. Strictly speaking, this violates our requirement that the encoder produce a sequence of channel symbols at the channel rate. This difficulty is circumvented by assuming that the encoder has a buffer into which the variable length codewords are put and out of which the channel takes its symbols at a fixed rate. If the channel

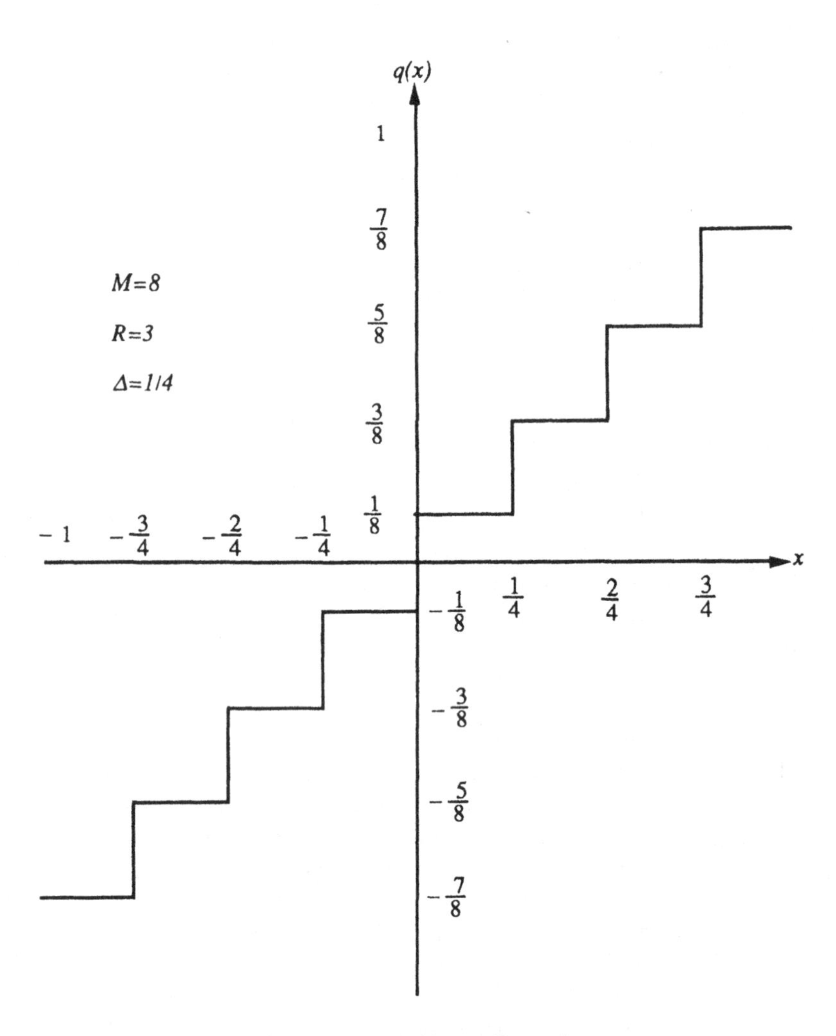

Figure 2.2: Uniform Quantizer

rate equals $\bar{l}R_{source}/k$, where $\bar{l}$ is the average length of the codewords, and if the source is such that successive k-dimensional vectors are stationary and ergodic, then on the average the channel will accept the codewords as they are produced. It will occasionally run out of symbols (underflow), during which time arbitrary symbols can be sent. If the buffer has a finite length, the encoder may occasionally lose symbols (overflow). Both events will have small probability if the buffer is large enough and if one inserts a coding delay of sufficient length. It should be apparent that variable-length codes can be difficult to handle theoretically and that they can cause increases in implementation complexity because of the required buffering. Hence one might wonder why they are used at all. There are two reasons. First, if one wishes perfectly noiseless data compression at rates near the theoretical limit, then variable-length codes must be used. (See, e.g., Gallager [28].) Second, variable length codes may provide more efficient implementations for certain data compression systems. For example, some data sources exhibit long periods of little change and short bursts of widely varying behavior. By sending fewer bits during the stable periods, one has more bits to devote to the active periods while keeping an overall average rate constant.

Other forms of variable-length codes are possible. Variable-length to fixed length codes are often used in noiseless coding (Tunstall codes, for example [77]) and variable-length to variable-length codes are also possible (e.g., arithmetic codes [77]).

A natural generalization of a memoryless vector coder is a vector coder with memory. Here one effectively has a different encoder for each input vector, where the encoder is selected in a manner that depends on the past. A fairly general model for such a system is that of a *sequential machine* or *feedback vector coder* or *state machine* or *recursive quantizer*: Suppose that we have a collection S of possible states, an initial state S_0, a next-state rule $g(\mathbf{x}, s)$ which assigns a new state $g(\mathbf{x}, s)$ for an input vector $\mathbf{x}$ and current state s, and a code mapping $f(\mathbf{x}, s)$ which assigns a channel vector $f(\mathbf{x}, s)$ when the input vector $\mathbf{x}$ is viewed when the encoder is in state s. The action of the encoder is then defined by the recursion relations

$$S_{n+1} = g((X_{nk}, \ldots, X_{(n+1)k-1}), S_n),$$

$$(U_{nK}, \ldots, U_{(n+1)K-1}) = f((X_{nk}, \ldots, X_{(n+1)k-1}), S_n).$$

A decoder can be defined similarly. As with memoryless vector coders, we say that the coder is a feedback scalar coder if $k = 1$.

An example of such a feedback scalar coder is a simple *delta modulator* as depicted in Figure 2.3. Suppose that the channel alphabet is ± 1 and the code dimension is $K = 1$. The state can be thought of as the last

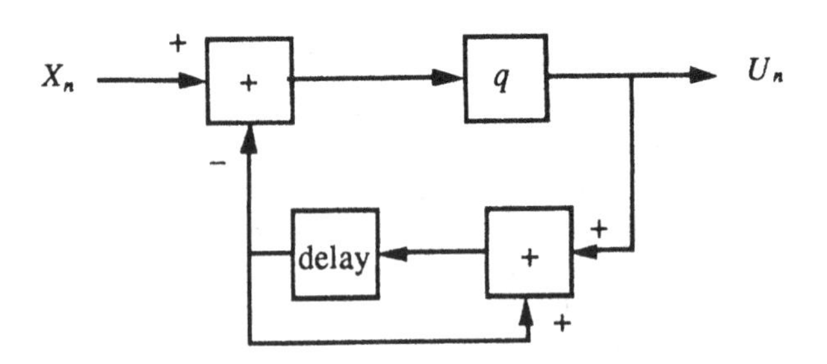

Encoder

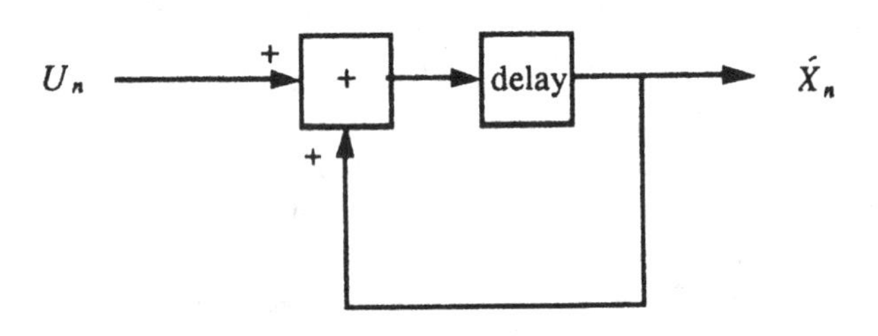

Decoder

Figure 2.3: Delta Modulator

reproduction value. The initial state is 0. Δ is a fixed parameter. Define the encoder function by

$$f(x, s) = \mathrm{sgn}(x - s),$$

where $\mathrm{sgn}(x)$ is the signum function or sign of x. Thus the encoder simply sends the sign of the "error" between the input and the current reproduction. Define the next state function by

$$g(x, s) = s + \Delta f(x, s)$$

and hence

$$S_{n+1} = S_n + \Delta U_n.$$

The decoder simply receives the U_n and puts out the state sequence using the above recursion and initial condition. Overall, the system tries to track the input sequence by a staircase function with steps of $\pm\Delta$. Note that this system has a very strong dependence on past inputs or outputs. Note further that we have constructed the next state function so that the decoder can track the encoder state knowing only the initial conditions and the channel symbols. We shall return to the Delta modulator in some detail in Chapter 6.

In the special case where there are only a finite number of states, the code is called a *finite state vector code*. Finite state vector codes are useful, for example, when one wants to use a finite collection of possible codes, each designed to be good when the source is exhibiting a certain behavior. The codes are selected according to recent behavior. Finite state codes were introduced by Shannon [73].

The delta-modulation example above also provides an example of another form of code: Instead of mapping nonoverlapping input vectors into codewords, we could map overlapping input vectors into code symbols. As an example, suppose that we have a mapping f of input k-vectors into channel symbols. Given the input sequence $\{X_n\}$ we then form the channel sequence $\{U_n\}$ by

$$U_n = f(X_n, X_{n-1}, \ldots, X_{n-k});$$

that is, each successive channel symbol is found by forming a function of a window of an input vector and then "sliding" the window one time unit to produce the next code symbol. This is an example of a *sliding-block code* or *sliding window code*. In the mathematics literature it is known as a *stationary code* or *shift-invariant code*. It can also be viewed as a time-invariant nonlinear filter. More generally, the code could depend on the infinite past, e.g.,

$$U_n = f(X_n, X_{n-1}, \ldots),$$

or on the infinite past and infinite future

$$U_n = f(\ldots, X_{n+1}, X_n, X_{n-1}, \ldots).$$

The code is said to have *delay* K if it has the form

$$U_n = f(X_{n+K}, \ldots, X_n, X_{n-1}, \ldots).$$

Any real world code is best modeled as having finite delay. It can be shown that if a stationary process is coded using a sliding-block code, then the resulting input/output process is jointly stationary.

Still more generally, the code may accept source vectors as inputs and produce channel vectors each time the window is slid. Both input or output vectors can have variable length. Defining such general codes, however, requires a lot of notation and adds little insight to the special case under consideration. One idea is, however, worth isolating: A general code will be said to have *finite delay* if there is an integer K such that for any n the encoded output at time n does not depend on inputs more than K samples into the future.

Most practical codes fit into the general cases cited above. Often they fit into more than one of the classes. Usually it is natural to assume that the encoder and decoder have the same structure, that is, they are both block codes or both sliding-block codes or both finite state machine codes. In some systems, however, different structures are used.

2.3 Code Rate

The first property of a general code that will prove important is its *rate*, a generalization of the rate or resolution of a quantizer and a concept closely related to the transmission rate of a channel. Roughly speaking, the rate of a code measures how rich the reproduction alphabet of a source code is, regardless of the particular structure of the code.

Suppose that we have a communications system with a noiseless channel with a transmission rate R_{trans} bits per source symbol and an input process or source $\{X_n\}$, which may be vector-valued. In order to maintain generality with a minimum of technical details, we make the simplifying assumption that the source is *one-sided*, that is, the random process $\{X_n;\ n = 0, 1, \ldots\}$ is defined only for nonnegative times. If the various codes depend on "past" values of their inputs, we assume that these values are set to some arbitrary values or initial conditions. This assumption corresponds to the physical notion of "turning on" the code at some precise time (chosen as time 0 for convenience), a reasonable assumption to make for any code designed and built by humans! It simplifies the mathematics by avoiding having to make the following arguments conditioned on past inputs, e.g., inputs before the code was turned on.

Suppose that we have a decoder which produces a reproduction process $\{\hat{X}_n;\ n = 0,1,2,\ldots\}$ from a channel output sequence $\{U_n;\ n = 0,1,2,\ldots\}$. For any integer N define M_N as the maximum number of possible output N-dimensional vectors $(\hat{X}_0, \hat{X}_1, \ldots, \hat{X}_{N-1})$ that can be produced by the system regardless of any future inputs to the decoder. We assume that this is a finite number. For example, suppose that the channel has transmission rate $R_{\text{trans}} = K/k \log ||B||$ bits per input symbol, where K channel symbols are produced in the same amount of time that k source (input) symbols are produced. Say that we have a block code with source and reproduction block lengths of N and hence the corresponding channel block length is $L = NK/k$, which we require to be an integer in order to ensure physical implementability. In this case the maximum number of N-dimensional outputs is the same as the maximum number of channel L-dimensional words or

$$M_N \leq ||B||^L = 2^{NR_{\text{trans}}}. \tag{2.3.1}$$

Furthermore observe that if the decoder provides a distinct reproduction word for each channel word, the above bound holds with equality for all N for which such block codes exist. Note that if the code instead had block length J for $J > N$, then all we could say in general would be that $M_N \leq 2^{JR}$ and hence we could have that $M_N > 2^{NR}$.

If the code is a binary Delta modulator, then $M_N = 2^N$ for any N, regardless of the initial state of the Delta modulator.

Define the Nth order *rate* R_N of the system to be $R_N = N^{-1} \log M_N$. The units of rate are bits per source symbol (if the logarithm is base two) since the time units of the reproduction vectors and the source vectors are assumed to be the same.

In the first example above, R_N is bounded above by R_{trans} and can be taken to be equal to R_{trans} if the decoder is chosen properly. In the second example of a block code of length J, we can only say that $R_N \geq R_{\text{trans}}$. In the Delta modulator example, $R_N = 1$.

For the asymptotic or long term performance results to come, it is of interest to characterize the Nth order resolution of a code for all large N. Toward this end we define the *limiting rate* or, simply, the *rate* of a code to be the smallest value R for which the following is true: Given any $\delta > 0$ there is a J large enough to ensure that $R_N \leq R + \delta$ for all $N \geq J$. (Mathematically, this is just the limit supremum $\limsup_{N\to\infty} R_N$.) For example, if the code is such that the limit $\lim_{N\to\infty} R_N$ exists, then this limit is the rate R of the code.

We next relate the rate of the decoder of a source coding system to the transmission rate of the channel for the practically important case of a finite delay code. Suppose that the decoder has a finite delay L in the following sense: For any output time n the reproduction symbol $\hat{X}_n$ is completely determined by the channel symbols that have occurred from time 0 up to the corresponding channel time (approximately $n\tau_s/\tau_c$) plus an additional $L\tau_s/\tau_c$ "future" channel symbols. In particular, we assume that there is a finite L such that $\hat{X}_n$ is a deterministic function of at most $(n+L)\tau_s/\tau_c$ channel symbols for all n. This condition will be met by all code structures modeling physically realizable codes. Thus for a fixed N, the entire reproduction vector $\hat{X}_0, \ldots, \hat{X}_{N-1}$ is determined by a sequence of $(N+L)\tau_s/\tau_c$ channel symbols. Since there are at most $||B||^{(N+L)\tau_s/\tau_c} = 2^{(N+L)R_{\text{trans}}}$ possible sequences of channel outputs in this time, there can be at most the same number of reproduction sequences, that is,

$$M_N \leq 2^{(N+L)R_{\text{trans}}}$$

and hence

$$R_N \leq \frac{N+L}{N} R_{\text{trans}}. \tag{2.3.2}$$

The limit of the right hand side as $N \to \infty$ yields R_{trans} which implies that for codes with finite (but arbitrarily large) delay

$$R \leq R_{\text{trans}}, \tag{2.3.3}$$

a fact that we formally state as a lemma:

Lemma 2.3.1 *The rate of a decoder with finite delay is bound above by the transmission rate of the channel, that is, Eq. (2.3.2) holds.*

In fact the finite delay assumption is not necessary, but it is general enough for our purposes and simpler to handle.

2.4 Code Performance

In order to have a theory which quantifies the tradeoffs between distortion or fidelity and communication rate, we must first provide a mathematical definition of distortion. Suppose that we have as an information source a random process $\{X_n\}$ with alphabet A, where in general A can be the real line or k-dimensional Euclidean space. Suppose that a communications system produces a reproduction process $\{\hat{X}_n\}$, where the $\hat{X}_n$ are drawn from an alphabet $\hat{A}$. Usually $\hat{A}$ is the same as A, but often it is convenient or necessary to consider different reproduction and input alphabets. A *distortion measure* d is an assignment of a nonnegative number $d(x, \hat{x})$ to every input symbol $x \in A$ and output symbol $\hat{x} \in \hat{A}$. In mathematical notation, a distortion measure is a function or mapping $d : A \times \hat{A} \to [0, \infty)$, where the notation means that d assigns a number in $[0, \infty) = \{r : 0 \leq r < \infty\}$ to every point in the cartesian product space $A \times \hat{A}$ consisting of all pairs $(x, \hat{x})$ with $x \in A$ and $\hat{x} \in \hat{A}$. The distortion $d(x, \hat{x})$ is intended to measure the distortion or cost incurred if an input x is reproduced as an output $\hat{x}$.

The selection of a distortion measure for a particular problem can be a difficult and often controversial problem. Ideally, the distortion should quantify to subjective quality, e.g., small distortion in an image coding system should mean a good looking image, large distortion a bad looking image. If one is to have a useful theory, the distortion measure should be amenable to mathematical analysis, e.g., one should be able to compute sums or integrals of the distortion and optimize such quantities via calculus. If the distortion measure is to be useful for practice, one should be able to actually measure it, that is, it should be computable. Unfortunately, one rarely has a distortion measure which has all three properties of subjective meaningfulness, mathematical tractability, and computability. In fact, the first property is often at odds with the second two since a distortion measure might need to be extremely complicated to even approximate subjective feelings of quality.

We consider several distortion measures on real vectors that have proved useful in theory or in practice. We will not treat the issue of how to actually choose a distortion measure for a particular application as that is more of a psychophysical testing issue than a data compression system design issue.

Perhaps the most common distortion measure in the literature is the squared error or mean squared error distortion: Suppose that A is k-dimensional Euclidean space $\mathcal{R}^k$ and that $\hat{A}$ is a subset of $\mathcal{R}^k$. Then the squared error distortion measure is defined as

$$d(\mathbf{x}, \hat{\mathbf{x}}) = \sum_{i=0}^{k-1} (x_i - \hat{x}_i)^2,$$

where $\mathbf{x} = (x_0, \ldots, x_{k-1})$. The squared error distortion measure is just the square of the Euclidean distance. It has the physical significance of being the instantaneous energy in the error if the error is viewed as a voltage across a 1 ohm resistor. Minimizing average error energy in a communications system is an intuitive and often tractable problem, but it is often inadequate as a subjectively meaningful distortion measure. For example, if the input process is sampled speech and the reproduction is formed by simply delaying the input, then the average squared error or *mean squared error (MSE)* will be quite large even though it sounds the same. The arithmetic mean of the squared-error can itself be considered a distortion measure, e.g., we can define

$$d(\mathbf{x}, \hat{\mathbf{x}}) = \frac{1}{k} \sum_{i=0}^{k-1} (x_i - \hat{x}_i)^2.$$

The Euclidean distance in $\mathcal{R}^k$ is an example of a *norm* and a *distance* or *metric* in Euclidean space. Both norms and distances provide other examples of distortion measures. In general, a *distance* or *metric* m on a real vector space is an assignment of a nonnegative number $m(\mathbf{x}, \hat{\mathbf{x}})$ to every pair of elements $\mathbf{x}$ and $\hat{\mathbf{x}}$ in the space so

1. $m(\mathbf{x}, \hat{\mathbf{x}}) = m(\hat{\mathbf{x}}, \mathbf{x})$ (symmetry);

2. $m(\mathbf{x},\hat{\mathbf{x}}) = 0$ if and only if $\mathbf{x} = \hat{\mathbf{x}}$;

3. for any vectors $\mathbf{x}$, $\hat{\mathbf{x}}$, $\mathbf{y}$,

$$m(\mathbf{x}, \hat{\mathbf{x}}) \leq m(\mathbf{x}, \mathbf{y}) + m(\mathbf{y}, \hat{\mathbf{x}}),$$

   that is, m satisfies a triangle inequality.

If the second property is not satisfied, than m is called a *pseudo-metric*.

A *norm* on a real vector space assigns a nonnegative number $||\mathbf{x}||$ to every vector in the space with the properties that

1. $||a\mathbf{x}|| = |a|||\mathbf{x}||$ for all real a,

2. $||\mathbf{x}|| = 0$ if and only if $\mathbf{x}$ is the vector of all 0's, and

3. $||\mathbf{x} + \mathbf{y}|| \leq ||\mathbf{x}|| + ||y||$; that is, norms satisfy the triangle inequality.

If the second condition is not satisfied, then $|\cdot|$ is called a *seminorm*.
The function

$$||\mathbf{x}|| = \left(\sum_{i=0}^{k-1} x_i^2 \right)^{1/2}$$

is a norm and

$$m(\mathbf{x}, \hat{\mathbf{x}}) = ||\mathbf{x} - \hat{\mathbf{x}}||$$

is a distance, where $\mathbf{x} - \hat{\mathbf{x}}$ denotes the vector whose coordinates are the differences of the coordinates of $\mathbf{x}$ and $\hat{\mathbf{x}}$. This distance is the well known Euclidean distance and hence the MSE is simply the square of the Euclidean distance.

Clearly any norm or distance on $\mathcal{R}^k$ provides a distortion measure, along with any nonnegative function such as a power of such norms and distances. We say that a distortion measure is a metric distortion measure if it is a metric, a norm distortion measure if it is the norm of the difference of the input and reproduction, and a norm-based distortion measure if it is some nonnegative function of the norm of the difference of its arguments. For example, the so-called l_p norms defined by

$$||\mathbf{x}||_p = (\sum_{i=0}^{k-1} x_i^p)^{1/p}$$

yield distances defined by

$$m(\mathbf{x}, \hat{\mathbf{x}}) = ||\mathbf{x} - \hat{\mathbf{x}}||_p.$$

Any nonnegative function of these distances is a valid distortion measure. These distortion measures provide an example of an important class of distortion measures: those that depend on their arguments only through their difference. A distortion measure d on real vector spaces is called a *difference distortion measure* if $d(\mathbf{x}, \hat{\mathbf{x}})$ depends on $\mathbf{x}$ and $\hat{\mathbf{x}}$ only through the difference $\mathbf{x} - \hat{\mathbf{x}}$.

Suppose that a $k \times k$ matrix $\mathbf{W} = \{W_{ij};\ i, j = 0, 1, \ldots, k-1\}$ is positive definite, that is, the quadratic form

$$\mathbf{x}^t \mathbf{W} \mathbf{x} = \sum_{i=0}^{k-1} \sum_{j=0}^{k-1} x_i W_{ij} x_j$$

is positive whenever $\mathbf{x}$ is not identically 0. Then the distortion measure defined by

$$d(\mathbf{x}, \hat{\mathbf{x}}) = (\mathbf{x} - \hat{\mathbf{x}})^t \mathbf{W} (\mathbf{x} - \hat{\mathbf{x}})$$

is called the *weighted squared error* or the *weighted quadratic* distortion measure. Such distortion measures allow one to weight certain vector components more heavily in computing the distortion. Note that this too is a difference distortion measure. The Mahalonobis distance of statistics is of this form.

An example of a distortion measure that is not a difference distortion is the following: Suppose that for each input vector $\mathbf{x}$ we have a different positive definite matrix $\mathbf{W}_x$. Then the distortion measure

$$d(\mathbf{x}, \hat{\mathbf{x}}) = (\mathbf{x} - \hat{\mathbf{x}})^t \mathbf{W}_{\mathbf{x}} (\mathbf{x} - \hat{\mathbf{x}})$$

is called an *input weighted squared error* distortion. For example, $\mathbf{W}$ might assign more weight to coordinates in which the input vector $\mathbf{x}$ had high (or low) energy. For instance, if $\mathbf{W}_x = 1/||\mathbf{x}||^2\mathbf{I}$, the identity matrix times the reciprocal of the energy in the vector, then $d(\mathbf{x}, \hat{\mathbf{x}})$ has the interpretation of being the short term noise to signal ratio and counts the noise energy more heavily when the signal energy is small.

A distortion measure of importance when the alphabets are discrete is the Hamming distortion measure defined by

$$d(\mathbf{x}, \hat{\mathbf{x}}) = \begin{cases} 0 & \text{if } \mathbf{x} = \hat{\mathbf{x}} \\ 1 & \text{otherwise;} \end{cases}$$

that is, there is no distortion if the symbol is perfectly reproduced and a fixed distortion if there is an error, no matter what the nature of the error.

Given we have a measure of the distortion of reproducing one input symbol for one output symbol, how do we compute the distortion of reproducing a sequence of input symbols as a given sequence of output symbols? In general, we could define a different distortion measure d_n for sequences of n input and output symbols (each of which might be a vector). Such a family of distortion measures $\{d_n;\ n = 1, 2, \ldots\}$, where $d_n(\mathbf{x}^n, \hat{\mathbf{x}}^n)$ is the distortion between an n dimensional input sequence $\mathbf{x}^n = (\mathbf{x}_0, \mathbf{x}_1, \ldots, \mathbf{x}_{n-1})$ with coordinates $\mathbf{x}_i$ in A and an n dimensional output sequence with coordinates in $\hat{A}$, is called a *fidelity criterion* in information theory. We shall emphasize a particular important special case: Suppose that we have a fidelity criterion yielding distortions $d_n(\mathbf{x}^n, \hat{\mathbf{x}}^n)$ for all n and all sequences $\mathbf{x}^n$ in A^n and $\hat{\mathbf{x}}^n$ in $\hat{A}^n$, respectively. We say that the fidelity criterion is *additive* or *single letter* if

$$d_n(\mathbf{x}^n, \hat{\mathbf{x}}^n) = \sum_{i=0}^{n-1} d_1(x_i, \hat{x}_i);$$

that is, the distortion between a sequence of n input symbols and n output symbols is simply the sum of the distortions incurred in each input/output symbol pair. This special case has the intuitive interpretation that the distortion between n successive symbols normalized by the number of symbols is simply the time average distortion for the sequence:

$$\frac{1}{n} d_n(\mathbf{x}^n, \hat{\mathbf{x}}^n) = \frac{1}{n} \sum_{i=0}^{n-1} d_1(x_i, \hat{x}_i).$$

Given such a fidelity criterion, the sample performance of a communications system with input process $\{X_n\}$ and reproduction process $\{\hat{X}_n\}$ can be measured by the long term time average distortion

$$\lim_{n\to\infty} \frac{1}{n} \sum_{i=0}^{n-1} d_1(X_i, \hat{X}_i),$$

if the limit exists, which it will if the pair process $\{X_n, \hat{X}_n\}$ is asymptotically mean stationary. We shall usually assume that the limit exists. In fact, if the source is asymptotically mean stationary, then fixed rate code structures considered all yield asymptotically mean stationary pair processes. (See, e.g., [42], [58], [55], [57], [41] for detailed treatments of the stationarity properties of sources, channels, and codes.) The limit will be a random variable, however, unless the pair process is also ergodic, in which case the limit is a constant. If the process is stationary, then the constant is the expectation of $d(X_0, \hat{X}_0)$.

It is much harder to prove ergodicity for specific code structures than it is to prove asymptotic mean stationarity. We have pointed out, however, that nonergodic stationary sources can be considered as a mixture of ergodic stationary sources. For this reason we will often assume ergodicity in derivations.

If the overall process $\{X_n, \hat{X}_n\}$ is asymptotically mean stationary and ergodic, then the limiting time average distortion is given by the limiting average of the expected distortion

$$\bar{d} = \bar{E}d_1(X_n, \hat{X}_n) = \lim_{n\to\infty} \frac{1}{n}\sum_{k=0}^{n-1} Ed_1(X_k, \hat{X}_k),$$

where the bar notation is consistent with the limiting averages defined for quasi-stationary processes. This suggests that the above limiting expectation is therefore a reasonable measure of the quality (or rather lack thereof) of the underlying system. Even when the system is not ergodic, this limiting expected distortion is the most common measure of performance. If the overall process $\{X_n, \hat{X}_n\}$ is stationary, then this simplifies to

$$\bar{d} = Ed_1(X_0, \hat{X}_0).$$

This happens if a stationary source is coded by a sliding-block code. Suppose that the overall process $\{X_n, \hat{X}_n\}$ is N-stationary for some integer N, that is, vectors of dimension N are stationary. This occurs, for example, if the original process is stationary and it is block coded by a block code of dimension N. In this case

$$\bar{d} = E\left(\frac{1}{N}\sum_{k=0}^{N-1} d_1(X_k, \hat{X}_k)\right) = \frac{1}{N}\sum_{k=0}^{N-1} E\left(d_1(X_k, \hat{X}_k)\right).$$

With a quantitative measure of the performance of a system, we can now formally define optimal systems.

2.5 Optimal Performance

Suppose that we are given a source $\{X_n\}$, a channel ν described by its alphabet and transmission rate, and a class of codes $\mathcal{G}$, where by a class of codes we mean the collection of all codes which satisfy some set of structural constraints, e.g., memoryless vector codes or finite-state vector codes. We assume that the original process, code structure, and channel are such that the overall source/reproduction process $\{X_n, \hat{X}_n\}$ is asymptotically mean stationary. We define the *optimal performance theoretically achievable* or *OPTA* function by

$$\Delta_X(\nu, \mathcal{G}) = \inf_{\text{codes in } \mathcal{G}} \bar{d}$$

If the source and code class are such that $\{X_n, \hat{X}_n\}$ is stationary, then

$$\Delta_X(\nu, \mathcal{G}) = \inf_{\text{codes in } \mathcal{G}} Ed_1(X_0, \hat{X}_0).$$

This is the case, for example, if a stationary source is coded with a sliding-block code.

If the source and code class are such that $\{X_n, \hat{X}_n\}$ is N-stationary, then

$$\Delta_X(\nu, \mathcal{G}) = \inf_{\text{codes in } \mathcal{G}} E\left(\frac{1}{N}\sum_{k=0}^{N-1} d_1(X_k, \hat{X}_k)\right).$$

This is the case, for example, if a stationary source is coded with a length N block code.

Ideally one would like to compute the OPTA function for arbitrary sources, channels, and code structures, but this is in general an impossible task. Hence we shall choose the code structures to be sufficiently general so that we can find or approximate or bound the OPTA function, but we may find that the resulting codes are impossibly complex to build. In the next chapters we develop OPTA functions for systems where they can be found explicitly, bounded, or approximated.

2.6 Information

One of the primary tools of information theory is a probabilistic definition of information which can be given operational significance in the form of bounds on optimal performance. While the measures of information are of importance and interest in their own right, we here consider them only as means towards an end and limit our development to the bare essentials. The interested reader can find complete developments of the basic properties of information measures in standard information theory texts such as

Gallager [28] and advanced (measure theoretic) treatments in Pinsker [69] and Gray[38].

The most important measure of information for our purposes is average mutual information: Suppose that X and Y are two random vectors. First suppose that X and Y are finite-valued and are described by a joint pmf $p_{X,Y}$ and let

$$p_X(x) = \sum_y p_{X,Y}(x,y) \text{ and } p_Y(y) = \sum_x p_{X,Y}(x,y)$$

be the induced marginals. The *average mutual information* between X and Y is defined by

$$I(X;Y) = \sum_{x,y} p_{X,Y}(x,y) \log \frac{p_{X,Y}(x,y)}{p_X(x)p_Y(y)}.$$

We define $0 \log 0 = 0$ and hence we can safely assume that the sum is only over those x and y for which the pmfs are strictly positive. If the logarithm is base two, then the units of information are called *bits*. If natural logarithms are considered, the units are called *nats*. Usually natural logarithms are more convenient for analysis and base 2 for intuition.

Next suppose that X and Y are not finite-valued but have either countably infinite or continuous alphabets. We can always form a finite approximation to the random vectors by using quantizers: Let q denote a mapping from the alphabet A_X of X into a finite alphabet and let r denote a mapping from the alphabet A_Y of Y into a finite alphabet. The random variables $q(X)$ and $r(Y)$ thus formed are finite valued and hence have an average mutual information defined as above. We now define average mutual information in the general case as

$$I(X;Y) = \sup_{q,r} I(q(X);r(Y)),$$

that is, the average mutual information between two random vectors is the supremum of the average mutual information achievable between quantized versions of the random vectors. The supremum is necessary as the maximum may not exist. If the random vectors have countably infinite alphabets, then the finite alphabet formula in terms of pmf's remains valid. If the random vectors are described by a joint pdf $f_{X,Y}$, then it can be shown that

$$I(X;Y) = \int dx \int dy f_{X,Y}(x,y) \log \frac{f_{X,Y}(x,y)}{f_X(x)f_Y(y)},$$

the integral analog of the pmf formula.

While the operational meaning of mutual information is supplied by the coding theorem and bounds to come, there are some interpretations which might help motivate its study. Since the discrete case is simpler and forms part of the definition for the continuous case, we often focus on finite-alphabet random variables. These properties are best introduced in terms of another information measure, the relative entropy between two pmf's: Suppose that we are given two pmf's p and q on a common space. The *relative entropy* (or *Kullback-Leibler number* or *discrimination* or *cross entropy*) of q with respect to p is defined by

$$H(p\|q) = \sum_k p(k) \log \frac{p(k)}{q(k)}$$

for pmf's and

$$H(f\|g) = \int dx f(x) \log \frac{f(x)}{g(x)}$$

for pdf's. The relative entropy can be viewed as a measure of the dissimilarity between two pmf's because of the following result.

Lemma 2.6.1 *The relative entropy between two pmf's is nonnegative. It is 0 if and only if the two pmf's are identical.*

Proof:

Consider natural logarithms. Using the elementary inequality $\ln x \leq x - 1$ with equality if and only if x is 1 implies that

$$-H(p\|q) = \sum_k p(k) \log \frac{q(k)}{p(k)} \leq \sum_k p(k) \left(\frac{q(k)}{p(k)} - 1 \right) = \sum_k q(k) - 1 = 0$$

with equality if and only if $p(k) = q(k)$ for all k. The proof for pdf's follows similarly.

Given that the relative entropy can be interpreted as a measure of dissimilarity between pmf's, then observing that

$$I(X;Y) = H(p_{X,Y} \| p_X p_Y)$$

points out that the average mutual information between two random variables can be viewed as a measure of how dissimilar their joint pmf is from what it would be if the two random variables were independent. Thus it is in a sense a measure of how much information the two random vectors provide about each other. It follows from the above lemma and formula that

$$I(X;Y) \geq 0$$

with equality if and only if the two random vectors are independent. Returning to the general case, the *entropy* or *average self information* of a random vector X is defined by

$$H(X) = I(X;X);$$

that is, the entropy of a random vector is the average mutual information it has about itself. In the discrete case where the random variable is specified by a pmf, this becomes

$$H(X) = -\sum_k p_X(k) \log p_X(k).$$

Since $0 < p_X(x) \leq 1$ for all x, we have that

$$H(X) \geq 0.$$

If the alphabet is finite with, say, K symbols, then application of the previous lemma with $p(x) = p_X(x)$ and $q(x) = 1/K$ yields

$$H(X) \leq \log K.$$

In the continuous case, the entropy is usually infinite. It is convenient, however, to give a name to the pdf analog of the above formula: If a random variable X is described by a pdf f_X, then the quantity

$$h(X) = -\int_x f_X(x) \log f_X(x)\, dx$$

is called the *differential entropy* of the random variable X. It should be emphasized that differential entropy does not behave like an entropy. For example, it can be negative or positive or infinite valued.

Mutual information can be decomposed into entropy terms. In the discrete alphabet case we have

$$I(X;Y) = H(X) + H(Y) - H(X,Y).$$

In the continuous case with pdf's this becomes

$$I(X;Y) = h(X) + h(Y) - h(X,Y).$$

Other useful decompositions follow from the definition of *conditional entropy*. Define the conditional entropy

$$H(X|Y) = H(X,Y) - H(Y) = -\sum_{x,y} p_{X,Y}(x,y) \log p_{X|Y}(x|y)$$

or

$$H(X|Y) = \sum_y p_Y(y) H(X|Y = y)$$

where

$$H(X|Y = y) = -\sum_x p_{X|Y}(x|y) \log p_{X|Y}(x|y)$$

is the entropy with respect to the conditional pmf $p_{X|Y}(.|y)$. With this notation,

$$I(X;Y) = H(X) - H(X|Y),$$

which has the interpretation that the mutual information between two random variables is the difference between the self information of one and the conditional information of one given the other variable. Similarly, by defining in the continuous case the conditional differential entropy

$$h(X|Y) = \int dy f_Y(y) h(X|Y = y),$$

where

$$h(X|Y = y) = -\int dx f_{X|Y}(x|y) \log f_{X|Y}(x|y),$$

we have that

$$I(X;Y) = h(X) - h(X|Y).$$

In some situations one has a combination of discrete and continuous random vectors or variables. Suppose that X is discrete and Y is continuous and can be described by a pdf f_Y. In this case we can write

$$I(X;Y) = H(X) - \int dy f_Y(y) H(X|Y = y), \tag{2.6.1}$$

where

$$H(X|Y = y) = -\sum_x p_{X|Y}(x|y) \log p_{X|Y}(x|y),$$

where $p_{X|Y}(x|y)$ is the conditional pmf for the discrete variable X given the continuous variable Y. Since $H(X|Y = y)$ is an ordinary entropy, it is nonnegative and hence

$$I(X;Y) \leq H(X). \tag{2.6.2}$$

2.6.1 Information and Entropy Rates

We will also be interested in the limiting behavior of average mutual information between vectors as the vector size becomes larger. Suppose that a process $\{X_n, Y_n\}$ is asymptotically mean stationary. It can be shown that under fairly general conditions the limiting average mutual information per symbol defined by

$$\bar{I}(X;Y) = \lim_{N\to\infty} \frac{1}{N} I(X^N;Y^N)$$

exists [69], [38]. For example, it exists if one of the alphabets is finite or if

$$\lim_{n\to\infty} I(X_0;X_1,\ldots,X_n) < \infty,$$

that is, the limiting information between the present and future is finite. This quantity is called the *information rate* of the input/output process. A special case of information rate is the entropy rate

$$\bar{H}(X) = \lim_{N\to\infty} \frac{1}{N} H(X^N),$$

where we consider the rate to be defined only if the limit exists. The limit will exist if the process is asymptotically mean stationary and the process has a discrete alphabet. It can be shown that if the process is stationary, then $N^{-1}H(X^N)$ is nonincreasing in N and hence

$$\bar{H}(X) = \inf_N \frac{H(X^N)}{N}$$

where the infimum is required since the minimum may not exist. If a process $\{X_n, Y_n\}$ is such that X has a discrete alphabet, then $H(X^n) \geq I(X^n;Y^n)$ implies that

$$\bar{H}(X) \geq \bar{I}(X;Y).$$

2.7 Limiting Properties

Information satisfies a form of ergodic theorem and this plays a key role in the proof of the source coding theorem. In this section we collect the appropriate definitions and state the prerequisite results without proof. Define the Nth order sample mutual information $i_N(X^N;Y^N)$ by

$$i_N(X^N;Y^N) = \begin{cases} \log \frac{f_{X^N,Y^N}(X^N,Y^N)}{f_{X^N}(X^N)f_{Y^N}(Y^N)} \\ \text{or} \\ \log \frac{p_{X^N,Y^N}(X^N,Y^N)}{p_{X^N}(X^N)p_{Y^N}(Y^N)} \end{cases}$$

depending on whether the random vectors are described by pmf's or pdf's. Note the potentially confusing dual use of the symbols X^N and Y^N: they are used in the subscripts to name the probability functions, but inside the parentheses they stand for the random vectors. Similarly, define for a discrete alphabet process $\{X_n\}$ the sample entropy $\eta_N(X^N) = i_N(X^N; X^N)$ given by

$$\eta_N(X^N) = -\log p_{X^N}(X^N).$$

Observe that

$$I(X^N; Y^N) = E i_N(X^N; Y^N)$$

and

$$H(X^N) = E\eta_N(X^N)$$

The sample entropy and sample information satisfy a form of an ergodic theorem. It can be shown for an asymptotically mean stationary and ergodic process with certain technical conditions that

$$\lim_{n\to\infty} \frac{1}{N}\eta_N(X^N) = \bar{H}(X).$$

If the convergence is in probability and the process is stationary, this is called the Shannon-McMillan theorem. If the convergence is with probability one and the process is stationary, the result is called the Shannon-McMillan-Brieman theorem. The result is also called the asymptotic equipartition property (AEP), the entropy theorem, and the ergodic theorem of information theory. More generally, if the process $\{X_n, Y_n\}$ is asymptotically mean stationary and satisfies some additional technical conditions, then

$$\lim_{N\to\infty} \frac{1}{N} i_N(X^N; Y^N) = \bar{I}(X; Y).$$

This result is a generalization of the Shannon-McMillan-Breiman theorem of information theory. Proofs may be found in Barron [3], Algoet and Cover [1], and Gray [38].

If the processes are iid, then the above results are ordinary ergodic theorems.

2.8 Related Reading

For the interested student, good treatments of information theory may be found in the books of Gallager [28], McEliece [66], and Blahut [7]. Extensive treatments of information and entropy rates and the convergence theorems of information theory may be found in Pinsker [69] and Gray [38].

Exercises

1. If d_1 and d_2 are metric (norm, norm-based) distortion measures and $\lambda \in (0,1)$, then is the distortion measure d defined by $d(x,\hat{x}) = \lambda d_1(x,\hat{x}) + (1-\lambda)d_2(x,\hat{x})$ also metric (norm, norm-based)?

2. Suppose that $\{X_n\}$ is an iid process. Show that if the process is discrete, then for all n

$$H(X_0,\ldots,X_{n-1}) = \sum_{i=0}^{n-1} H(X_i).$$

   Prove the corresponding result for differential entropy. Show that in the discrete iid case $\bar{H}(X) = H(X_0)$.

3. Evaluate or bound $H(X)$ for the pmf's of Chapter 2.

4. Evaluate the differential entropy $h(X)$ for a Gaussian random variable and a uniform random variable. Give an example of a random variable with a negative differential entropy.

5. If X is a discrete random variable with N letters, what is its maximum possible entropy and what pmf yields the maximum? If X is a continuous 0 mean random variable with $E(X^2) = \sigma^2$, what is its maximum possible differential entropy and what pdf yields the maximum?

6. Suppose that X and Y are two continuous real-valued random variables. Is it true that

$$h(X) \geq h(X|Y) \geq 0.$$

   Show that

$$h(X-Y|Y) = h(X|Y)$$

   and

$$h(X|Y) \leq h(X-Y).$$

7. Prove the following entropy relations (all random variables are discrete):

   (a)

$$H(X|Y) \geq H(X|Y,Z)$$

(b)

$$H(X^N|Y) \leq \sum_{k=0}^{N-1} H(X_k|Y)$$

(c) If $\{X_n, Y_n\}$ is a process for which $\{X_n\}$ is iid, then

$$I(X^N; Y^N) \geq \sum_{k=0}^{N-1} I(X_k; Y_k).$$

If the process $\{X_n, Y_n\}$ is iid, then the above inequality is an equality.

8. Simulate encoding a uniformly distributed random variable using a uniform quantizer having $M = 2, 4, 8$ levels. Find the average squared error and the entropy of the quantized signal.

Chapter 3

Distortion–Rate Theory

3.1 Introduction

Distortion-rate theory is the information theoretic approach to the study of optimal source coding systems, including systems for quantization and data compression. It is also called rate-distortion theory and the theory of source coding with a fidelity criterion. The approach was first suggested in Shannon's original development of information theory [73] and it was fully developed for memoryless and Markov sources in [74]. It was subsequently generalized to stationary ergodic processes with discrete alphabets and to Gaussian processes by Gallager [28] and to stationary and abstract alphabets by Berger [5] (as corrected by Dunham [26]). Elementary treatments may also be found in [79] and [66].

As with the more famous channel coding theory of Shannon, the source coding theorem combines the ergodic theorem with the idea of *random coding* in order to characterize the optimal achievable behavior of a particular mathematical model of a communication system. This behavior is characterized by the *distortion-rate function* (or *rate-distortion function*) of the source and distortion measure. In the next section we introduce the distortion-rate function, which we will abbreviate to DRF, and some of its basic properties. A natural way to introduce the DRF is within a development of the easy half of the Shannon source coding theorem, which says that no source coding system can yield average distortion less than the DRF. Thus Shannon theory immediately provides an unbeatable bound for comparison with real systems. The proof of this result is surprisingly easy in principle: One replaces the deterministic source codes used in practice by random mappings and than optimizes over all such mappings meeting a constraint (involving mutual information) which is satisfied by the deterministic codes. This optimum over random mappings is just the DRF. Since the class of random mappings includes the deterministic codes, no deterministic code can perform any better than the DRF.

The fact that the DRF is actually achievable by deterministic codes in a limiting sense is called the positive source coding theorem, and this result is deferred to Sections 3.3 and 3.4 where it is proved for two special cases.

3.2 Distortion-Rate Functions

In this section we develop the Shannon distortion-rate function (DRF) as a lower bound to the achievable performance of a communications system. The basic idea of the bound is simple: instead of optimizing over all codes of some given structure subject to a transmission rate constraint, we permit random mappings (that is, channels or conditional probabilities instead of deterministic mappings) and constrain the information rate. We shall see that this is a direct generalization of the class of codes and hence optimizing performance over the larger class must yield a lower bound to the performance obtainable in the smaller class of deterministic codes.

Suppose that we have a communications system with a noiseless channel with a transmission rate R_{trans} bits per source symbol and an input process or source $\{X_n\}$, which may be vector-valued. Suppose that we have an encoder and decoder which produce a final reproduction process $\{\hat{X}_n\}$. For any dimension N we can consider the Nth order average distortion between an input N-tuple $X^N = (X_0, X_1, \ldots, X_{N-1})$ and the corresponding reproduction vector $\hat{X}^N = (\hat{X}_0, \hat{X}_1, \ldots, \hat{X}_{N-1})$. Since the system is a code, in principle $\hat{X}^N$ is a deterministic function of the input (and possibly an initial code state), but it may depend on inputs into the future. Thus we could write the average distortion as something like $\bar{d}_N = E(d_N(X^N, \hat{X}^N(X^L)))$, for some possibly large (but finite) L. From the fundamental theorem of expectation, however, we could find the distribution of the random vector, say $X^N, \hat{X}^N$, and use this distribution to compute the average distortion, that is, to write

$$\bar{d}_N = E(d_N(X^N, \hat{X}^N)).$$

An expression of this form holds for block codes, sliding-block codes, and recursive codes, but the distribution may be a complicated function of the source distribution and the code structure. If the distortion measure is additive, then we have for the given code that

$$Ed_N(X^N, \hat{X}^N) = E\left(\sum_{k=0}^{N-1} d_1(X_k, \hat{X}_k)\right) = \sum_{k=0}^{N-1} Ed_1(X_k, \hat{X}_k).$$

Suppose that the Nth order rate of the code and channel is R_N. Since this is the Nth order average distortion for a particular code, it clearly must be no smaller than the smallest average distortion achievable by *any* code of the same rate. Since a deterministic mapping is a special case of a random mapping, this in turn must be no smaller than the smallest average distortion achievable using any conditional probability measure for the reproduction $\hat{X}^N$ given the input vector X^N, where the collection of possible outputs is assumed to be the same as that of the given code. Since the rate is R_N bits per source symbol, the maximum number of possible values that the reproduction vector can assume is $M_N = 2^{NR_N}$ and hence the entropy of the random reproduction vector $\hat{X}^N$ satisfies the bound

$$H(\hat{X}^N) \leq \log M_N = NR_N$$

and hence

$$I(X^N; \hat{X}^N) \leq H(\hat{X}^N) \leq NR_N. \tag{3.2.1}$$

Since the average mutual information between the input vector and reproduction vector is no greater than NR_N, the performance of the given system can be no better than that obtainable by the best random mapping (as before), where now we do not require a finite alphabet for the reproduction, but we do require that the average mutual information between the input and the reproduction be less than NR_N. Summarizing this development we have the following lower bound to the Nth order distortion for any system with decoder rate R_N:

$$\bar{d}_N \geq \inf Ed_N(X^N, Y^N),$$

where the infimum is over all conditional probability distributions for Y^N given X^N such that the resulting mutual information satisfies

$$\frac{1}{N} I(X^N; Y^N) \leq R_N.$$

In order to give the bound a name we now formally define the Nth order *distortion-rate function* by

$$D_N(R) = \inf Ed_N(X^N, Y^N), \tag{3.2.2}$$

where the infimum is over all conditional probability distributions for Y^N given X^N such that the resulting mutual information satisfies

$$\frac{1}{N} I(X^N; Y^N) \leq R. \tag{3.2.3}$$

To be more concrete, suppose that the input vector is continuous and described by a pdf f_{X^N}. Then

$$D_N(R) = \inf_{f_{Y^N|X^N}} \int dx \int dy f_{X^N}(x) f_{Y^N|X^N}(y|x) d_N(x, y),$$

where the infimum is over all conditional pdf's $f_{Y^N|X^N}$ such that

$$\int dx \int dy f_{X^N}(x) f_{Y^N|X^N}(y|x) \log \frac{f_{X^N}(x) f_{Y^N|X^N}(y|x)}{f_{X^N}(x) f_{Y^N}(y)} \leq NR.$$

If the input is discrete, then a similar formula follows for pmf's:

$$D_N(R) = \inf_{p_{Y^N|X^N}} \sum_x \sum_y p_{X^N}(x) p_{Y^N|X^N}(y|x) d_N(X^N, Y^N).$$

where the infimum is over all conditional pmf's such that $p_{Y^N|X^N}$ such that

$$\sum_{x,y} p_{X^N}(x) p_{Y^N|X^N}(y|x) \log \frac{p_{X^N}(x) p_{Y^N|X^N}(y|x)}{p_{X^N}(x) p_{Y^N}(y)} \leq NR.$$

In the discrete case it can be shown that the infimum is actually a minimum.

We summarize the above development in a lemma:

Lemma 3.2.1 *Given a source, channel, and code, if for an integer N the rate of the decoder is R_N and the average distortion of the source, code, and channel is $\bar{d}_N$, then*

$$\bar{d}_N \geq D_N(R_N),$$

where $D_N(R)$ is the Nth order distortion-rate function of the source and distortion measure.

Before continuing it is useful to point out an elementary property of distortion-rate functions.

Lemma 3.2.2 *$D_N(R)$ is nonincreasing in R, that is, if $R_1 \geq R_2$, then $D_N(R_1) \leq D_N(R_2)$.*

Proof: The minimization defining $D_N(R_1)$ is over a collection of conditional distributions which includes that defining $D(R_2)$, hence the resulting infimum can be no larger than that achievable over the smaller collection.

The simple bound of Lemma 3.2.1 is not always immediately useful because it is in terms of the rate of the decoder rather than the transmission rate of the channel and the rate can vary with N. If we happen to be considering the special case of a block encoder and decoder with common source and reproduction block length N and a corresponding channel block length of, say, K, then the channel transmission rate R is $K/N \log ||B||$ bits per source symbol, then $M_N = 2^{NR}$ and $R_N = R$. More generally, we remove the dependence of the above bound on N by considering the rate and the long term average distortion as in the previous chapter. Recall that for a given system,

$$\bar{d} = \lim_{N\to\infty} \frac{1}{N}\bar{d}_N$$

is the limiting per symbol average distortion. This immediately gives us the bound

$$\bar{d} \geq \lim_{N\to\infty} \frac{1}{N} D_N(R_N),$$

provided the limit exists. From (2.3.2) given $\delta > 0$ we can choose an N_0 large enough so that $R_N \leq R + \delta$ for all $N \geq N_0$, where R is the rate of the decoder. Furthermore, from Lemma 2.3.1 this bound still holds if R is considered as the transmission rate of the channel since the rate is bound above by the transmission rate. Thus for large N, $D_N(R_N) \geq D_N(R+\delta)$ from Lemma 3.2.2. We therefore have that for a transmission rate of R and any $\delta > 0$, no matter how small, that

$$\bar{d} \geq \lim_{N\to\infty} \frac{1}{N} D_N(R+\delta).$$

To formally state the main result of this section requires only two additional details. First we give a name to the above limit and define the ***distortion-rate function*** of the source and distortion measure to be the limit of the Nth order distortion-rate functions per source symbol as $N \to \infty$:

$$D(R) = \lim_{N\to\infty} \frac{1}{N} D_N(R). \tag{3.2.4}$$

If the limit does not exist the distortion-rate function is not defined. Second, we require a result proved later: Provided $R > 0$, $D(R)$ is a continuous function of R. Thus given any $\epsilon > 0$ we can find a $\delta > 0$ sufficiently small to ensure that

$$D(R+\delta) \geq D(R) - \epsilon. \tag{3.2.5}$$

Putting these facts together yields the following basic bound:

Theorem 3.2.1 *Given source and distortion measure for which the distortion-rate function $D(R)$ is well defined and an encoder/channel/decoder for which the channel has transmission rate R, the decoder has finite delay, and the average distortion $\bar{d}$ is well defined, then*

$$\bar{d} \geq D(R);$$

that is, no encoder and decoder can yield lower average distortion than the distortion-rate function evaluated at the transmission rate. Thus the OPTA function satisfies

$$\Delta(\nu, \mathcal{G}) \geq D(R)$$

for any channel ν with transmission rate R and for any class of codes $\mathcal{G}$ with finite delay decoders for which $\bar{d}$ is well defined.

This result is called a *negative coding theorem* or *converse coding theorem* because it states that no source coding system can yield performance better than that given by the distortion-rate function.

An immediate question is whether the bound is useful. In particular, can it be computed for interesting sources and distortion measures and is it a realistic bound, that is, is it a good bound? In a later chapter we will see examples where $D(R)$ can be found analytically, evaluated numerically, or approximated. We first consider the second question and show that not only is $D(R)$ a lower bound to the achievable performance, it is a bound that can in principle be almost achieved for some sources and code classes. The caveat "in principle" is required because the code complexity required to achieve performance near $D(R)$ may not be practicable.

3.3 Almost Noiseless Codes

Before turning to the positive coding theorems showing that performance arbitrarily close to the DRF can be achieved by a variety of coding structures, we consider a very special case. This provides a result of interest in its own right and a warmup for the more complicated result. Suppose that the original source is a discrete alphabet stationary and ergodic source. (The result generalizes to asymptotically mean stationary and ergodic sources.) Suppose that the Hamming distortion measure is used as the per-letter distortion in a single-letter fidelity criterion. Then for a block code of length N

$$\frac{1}{N} E d_N(X^N; \hat{X}^N) = \frac{1}{N} \sum_{k=0}^{N-1} \Pr(X_k \neq \hat{X}_k), \tag{3.3.1}$$

the average probability of symbol error. Suppose that we are told that the transmission rate is R and that $R > \bar{H}(X)$. What is the OPTA for block codes and what is the distortion-rate function? We shall show that the answer to both questions is 0; that is, the given rate is sufficient to code with arbitrarily small distortion. Codes yielding very small distortion are called "almost noiseless." Actually achieving 0 distortion is the goal of noiseless codes as treated, for example, in Chapter 9 of [30].

Define $\delta = (R - \bar{H}(X)) > 0$, by assumption. For each N define the set of "good" N-dimensional vectors G_N by

$$G_N = \{x^N : \frac{1}{N}\eta_N(X^N) \leq \bar{H}(X) + \delta/2\}.$$

These sequences are good in the sense that their sample entropy cannot be much larger than the limiting entropy rate. We will use the collection G_N as a block code. Note that for each vector x^N in G_N

$$p_{X^N}(x^N) \geq 2^{-N(\bar{H}(X)+\delta/2)},$$

and hence since this is the smallest probability of any good vector,

$$1 \geq P(G_N) = \sum_{x^N \in G_N} p_{X^N}(x^N) \geq \sum_{x^N \in G_N} 2^{-N(\bar{H}(X)+\delta/2)}$$

$$= ||G_N|| 2^{-N(\bar{H}(X)+\delta/2)},$$

where $||G_N||$ denotes the number of vectors in G_N. Thus we have that

$$||G_N|| \leq 2^{N(\bar{H}(X)+\delta/2)} = 2^{N(R-\delta/2)}.$$

If a channel has transmission rate $R = K/k \log ||B||$, where B is the channel alphabet and where K channel symbols are produced in the same time as k source symbols, then there exist block codes mapping input blocks of length N into channel blocks of length $L = KN/k$ for all N for which L is an integer. If the decoder yields distinct output vectors for each distinct channel L-tuple, then the rate of the decoder is exactly R. Thus there are 2^{NR} possible decoder words and corresponding channel words and we can therefore assign to each of the words in G_N a channel vector and have at least one channel vector left over, which we call the garbage vector. (Actually, we could use one of the regular channel vectors as the garbage vector, but it clarifies things a bit if it is separated out.)

The encoder works as follows: Given a source vector x^N, see if it is in G_N. If so, send the corresponding channel vector. If not, send the garbage vector. The source word will be communicated perfectly if it is in G_N. If it is not in G_N, then at worst every symbol would be in error. Thus the average distortion is bounded above by

$$\frac{1}{N}\bar{d}_N = \frac{1}{N}\sum_{k=0}^{N-1} P(X_k \neq \hat{X}_k)$$

$$= \frac{1}{N}\sum_{k=0}^{N-1}[P(X_k \neq \hat{X}_k|X^N \in G_N)P(X^N \in G_N)$$

$$+P(X_k \neq \hat{X}_k|X^N \notin G_N)P(X^N \notin G_N)]$$

$$\leq \frac{1}{N}\sum_{k=0}^{N-1} P(X^N \in G_N^c) = P(X^N \in G_N^c)$$

$$= P(\frac{1}{N}\eta_N(X^N) > \bar{H}(X) + \delta/2).$$

The Shannon-McMillan theorem guarantees that $\eta_N(X^N)/N$ converges in probability to $\bar{H}(X)$ and hence that the above probability goes to 0 as $N \to \infty$. Thus given any $\epsilon > 0$, we can find an N and a block code of length N and transmission rate R such that $\bar{d}_N \leq \epsilon$ provided only that $R > \bar{H}(X)$. Since this is true for any ϵ, the OPTA for this class is 0. Since the distortion-rate function is smaller than the OPTA, it too must be 0. Observe, however, that very large blocks may be required. We summarize these results as a lemma.

Lemma 3.3.1 *Suppose that $\{X_n\}$ is an ergodic stationary discrete alphabet source and d_n is a single-letter fidelity criterion with a Hamming per letter distortion. Let ν be a noiseless channel with transmission rate R and $\mathcal{G}$ the class of all rate R block codes. If $R > \bar{H}(X)$ then*

$$\Delta(\nu, \mathcal{G}) = D(R) = 0.$$

The above case is special because the alphabet is discrete, a special distortion measure is used, and only large rates are considered. It is important, however, for several reasons. First, it provides operational significance for the entropy rate of a source: It is the transmission rate needed to communicate a source almost noiselessly using block codes (and hence also more generally with feedback or recursive codes). As an example, a source with an alphabet of four symbols but with an entropy rate of only 1 bit per symbol can be communicated through a binary channel with one source symbol for each channel symbol and the average probability of error can be made arbitrarily small by using long enough codes. Thus one has compressed the original two bits per sample down to one bit per sample. There will, however, be occasional errors. Second, it demonstrates an example of the role played by the ergodic theorem in proving coding theorems. Here we were able to provide a fairly simple construction for the code: We effectively keep all the most likely source vectors and throw away the rest. The general construction will unfortunately be more complicated, but again the ergodic theorem will be crucial to the proof.

3.4 The Source Coding Theorem for Block Codes

In this section we show that performance arbitrarily near the distortion-rate function is achievable using block codes or vector quantizers. Since a block code is a special case of a finite-state code (with only one state), it also proves that performance near the DRF is possible with the more general class. The main distinction between the classes is that of implementation complexity: codes of either kind can yield near optimal performance, but one structure may be much simpler to implement for a given performance and rate.

Unfortunately, these so-called positive coding theorems are much harder to prove in the general case then was the negative coding theorem. In order to prove reasonably general positive coding theorems with a minimum of unenlightening technical details, we make some assumptions and we state without proof a useful property of DRFs. The assumptions are that (1) the source is stationary and ergodic, and (2) the fidelity criterion is additive and the distortion measure is bounded in the sense that there is a finite $d_{\max}$ such that $d_1(x,y) \leq d_{\max}$ for all x, y. Positive source coding theorems can be proved more generally for asymptotically mean stationary sources without requiring ergodicity, but the proof is highly technical[38],[56]. We point out here simply that it can be shown that the OPTA function of an asymptotically mean stationary random process is the same as the OPTA of a related stationary source and hence one can assume stationarity without any real loss of generality. Nonergodic sources are handled using a trick from ergodic theory called the *ergodic decomposition* which states that any stationary nonergodic source can be decomposed into a mixture of ergodic sources and that the OPTA function has a similar breakup. The bounded assumption is unfortunate since it is not satisfied by the common mean squared error, but it greatly simplifies the proof and generalizations can be found in the literature by the persistent.

The required property of a DRF is the following: If a source $\{X_n\}$ is stationary and ergodic and has a DRF $D(R)$, then for any $R > 0$ and $\delta > 0$ there is a stationary and ergodic process $\{X_n, Y_n\}$ such that

$$Ed_1(X_0, Y_0) \leq D(R) + \delta \tag{3.4.1}$$

and

$$\bar{I}(X;Y) \leq R; \tag{3.4.2}$$

that is, there is a random process with behavior similar to the random vectors giving the Nth order DRF's.

This result is a property of the so-called process definition of a distortion-rate function. The result was first proved by using the block source coding theorem by Marton [62] and Gray, Neuhoff, and Omura [43]. A direct proof not assuming the coding theorem may be found in [38]. The direct proof is preferable in the current approach in order to avoid the circular argument of using a result based on the coding theorem to prove the coding theorem. The general proof of this result is straightforward but tedious. It is not too hard to show, however, that it is true for iid sources, in which case the $\{X_n, Y_n\}$ process is also iid This case is developed in the exercises.

3.4.1 Block Codes

A block code consists of an encoder mapping of blocks of, say, N source symbols into a block of, say, K channel symbols and a decoder mapping of the channel block into a block of reproduction symbols. If the transmission rate of the channel is $R = K/N \log ||B||$, then the maximum Nth order rate is R since there are at most $||B||^K = 2^{NR}$ possible channel blocks producing the reproduction blocks. Furthermore, the block code will have Nth order rate of exactly R if the decoder assigns distinct reproduction words to all possible channel blocks. The collection of possible reproduction vectors is called a *code book*. Since we have an optimality criterion of minimizing the average distortion, given the code book no encoder can do better than that which views the source vector and then produces the channel vector which decodes into the reproduction vector with the minimum possible distortion. This observation generalizes the basic property of scalar and vector quantizers that the best encoders are minimum distortion mappings. Because of its importance, we formalize it as a lemma:

Lemma 3.4.1 *Suppose that a block source code has an encoder α mapping source vectors $\mathbf{x}$ into channel vectors $\mathbf{u} = \alpha(\mathbf{x})$ and a decoder β mapping channel vectors $\mathbf{u}$ into reproduction vectors $\hat{\mathbf{x}} = \beta(\mathbf{u})$. Let $\mathcal{W}$ be the code book $\{\beta(\mathbf{u});$ all $\mathbf{u}\}$. Then the encoder that is optimal in the sense of minimizing*

$$E(d_N(X^N, \beta(\alpha(X^N))))$$

over all encoders α is given by

$$\alpha^*(\mathbf{x}) = \min_{\mathbf{u}}{}^{-1} d(\mathbf{x}, \beta(\mathbf{u})), \tag{3.4.3}$$

where the inverse minimum notation means that the right hand side is that $\mathbf{u}$ yielding the given minimum. Thus

$$d(\mathbf{x}, \beta(\alpha^*(\mathbf{x})) = \min_{\hat{x} \in \mathcal{W}} d(\mathbf{x}, \hat{\mathbf{x}});$$

that is, the optimal encoder is a minimum distortion or nearest neighbor mapping.

A block source code with a minimum distortion mapping is commonly called a *vector quantizer*.

We can therefore consider a block code to be uniquely described by the code book: The encoder finds the best reproduction vector in the code book, i.e., the one with the minimum distortion. The index of this codeword is communicated over the noiseless channel to the receiver, which then produces the output using a simple table lookup.

3.4.2 A Coding Theorem

We now have the material needed to state and prove a positive coding theorem for block codes. We consider the case of continuous random variables described by pdf's. The discrete result follows in a similar manner via pmf's. The principal results are given as a theorem containing the fundamental result and two immediate corollaries. The remainder of the section is then devoted to the proof of the theorem.

Theorem 3.4.1 *Let $\{X_n\}$ be a stationary and ergodic source and let ν be a noiseless channel with transmission rate R. Let $\{d_n\}$ be a bounded single-letter fidelity criterion, and let $D(R)$ be the distortion-rate function. Given arbitrarily small numbers $\epsilon/2 > \delta > 0$ there exists for N sufficiently large a block length N source code with*

$$\bar{d} = \frac{1}{N}\bar{d}_N \leq D(R-\epsilon) + \delta.$$

In other words, there exists a block code with distortion very close to the distortion-rate function evaluated at a slightly smaller rate than the available transmission rate.

The theorem and the continuity of the DRF yield the similar statement given in the following corollary:

Corollary 3.4.1 *Given the conditions of the theorem and an arbitrarily small number $\delta > 0$ there exists for N sufficiently large a block length N source code with*

$$\bar{d} = \frac{1}{N}\bar{d}_N \leq D(R) + \delta.$$

That is, performance arbitrarily close to the DRF at rate R can be achieved. This follows because making ϵ small enough makes $D(R-\epsilon)$ as close as desired to $D(R)$.

Corollary 3.4.2 *Let the source and channel be as in the theorem. Let $\mathcal{G}$ denote any class of codes containing the block codes. Then*

$$\Delta(\nu,\mathcal{G}) \leq D(R).$$

This corollary follows from the fact that block codes yield performance arbitrarily close to the DRF, hence the optimal performance over all such codes must be no greater than the DRF.

Combining the preceeding corollary and Theorem 3.2.1 provides a characterization of the OPTA function for the class considered.

Corollary 3.4.3 *Given a stationary and ergodic source and a bounded single-letter fidelity criterion and a noiseless channel with transmission rate R, let $\mathcal{G}$ denote any class of codes containing block codes of all possible lengths. Then*

$$\Delta(\nu,\mathcal{G}) = D(R).$$

While Theorem 3.2.1 shows that $D(R)$ provides a lower bound to the OPTA for other code classes such as sliding-block codes, we thus far have a positive coding theorem only for block codes and their generalizations.

Proof of the Theorem:

First observe that that $N^{-1}\bar{d}_N = \bar{d}$ for a length N block code and a stationary source. Hence we need to show that $N^{-1}\bar{d}_N$ can be made close to $D(R-\epsilon)$.

The fundamental idea of Shannon's source coding theorem is this: for a fixed block size N, choose a code at random according to a distribution implied by the distortion-rate function. That is, perform approximately $M_N = 2^{NR}$ independent random selections of blocks of length N to form a code book. This code book is then used to encode the source using a minimum distortion mapping as above. We compute the average distortion over this double-random experiment (random code book selection followed by use of the chosen code to encode the random source). We will find that if the code generation distribution is properly chosen, then this average will be near $D(R)$. If the average over all randomly selected codes is near $D(R)$, however, then there must be at least one code such that the average distortion over the source distribution for that one code is also near $D(R)$. In other words, if an average is near a value, some of the terms being averaged must also be near that value. This means that there exists at least one code with performance near $D(R)$. Unfortunately the proof only demonstrates the existence of such codes, it does not show how to construct them.

To find the distribution for generating the random codes we use the stationary and ergodic process of equations (3.4.1)–(3.4.2) for $D(R-\epsilon)$, that is, we assume that $\{X_n, Y_n\}$ is a stationary and ergodic process and that

$$Ed_1(X_0, Y_0) \leq D(R - \epsilon) + \delta/2 \tag{3.4.4}$$

and

$$\bar{I}(X;Y) \leq R - \epsilon. \tag{3.4.5}$$

Choose N large; exactly how large will be specified later. Denote the vector pdf's for (X^N, Y^N), X^N, and Y^N by f_{X^N,Y^N}, f_{X^N}, and f_{Y^N}.

For any N we can generate a code book $\mathcal{W}$ at random according to f_{Y^N} as described above. To be precise, consider the random code book as a large random vector $W_1, \ldots, W_M$, where

$$M = 2^{N(R-\epsilon/4)}.$$

For simplicity we assume this to be an integer (otherwise just use the largest integer contained in M). Since the transmission rate of the channel is R, it can support codes of rate R for all those blocklengths N for which there exist block codes mapping N source symbols into L channel symbols, where $L = KN/k$ and K channel symbols are produced in the same time as k source symbols. Thus we are guaranteed that the block code decoder described by the above code book which has rate less than R can be communicated using the L symbols and hence the decoder can be informed which of the M reproduction words was selected by the encoder. Note that we choose M to have rate slightly less than R but slightly more than the rate used in (3.4.5).

The W_n are independent random vectors and the marginal pdfs for the W_n are given by f_{Y^N}. Thus the pdf for the randomly selected code can be expressed as

$$f_{\mathcal{W}}(\mathbf{w}) = f_{\mathcal{W}}(w_1, \ldots, w_M) = \prod_{i=1}^{M} f_{Y^N}(w_i).$$

This code book is then used with the optimal encoder and we denote the resulting average distortion (over code book generation and the source) by

$$\bar{\rho}_N = E\bar{\rho}_N(\mathbf{w}) = \int d\mathbf{w} f_{\mathcal{W}}(\mathbf{w}) \bar{\rho}_N(\mathbf{w})$$

where

$$\bar{\rho}_N(\mathbf{w}) = Ed_N(X^N, \mathbf{w}) = \int dx^N f_{X^N}(x^N) d_N(x^N, \mathbf{w}),$$

where

$$d_N(x^N, \mathbf{w}) = \min_{y \in \mathbf{w}} d_N(X^N, y).$$

Summarizing the above we have that

$$\bar{\rho}_N = \int d\mathbf{w} f_{\mathcal{W}}(\mathbf{w}) \int dx^N f_{X^N}(x^N) d_N(x^N, \mathbf{w})$$

$$= \int dx^N f_{X^N}(x^N) [\int d\mathbf{w} f_{\mathcal{W}}(\mathbf{w}) d_N(x^N, \mathbf{w})].$$

Abbreviate $D(R-\epsilon)$ to D and break up the integral over x into two pieces:

$$\int dx^N f_{X^N}(x^N) [\int d\mathbf{w} f_{\mathcal{W}}(\mathbf{w}) d_N(x^N, \mathbf{w})]$$

$$= \int dx^N f_{X^N}(x^N) [\int_{\mathbf{w}: N^{-1} d_N(x^N, \mathbf{w}) \leq D + \delta/2} d\mathbf{w} f_{\mathcal{W}}(\mathbf{w}) d_N(x^N, \mathbf{w})$$

$$+ \int_{\mathbf{w}: N^{-1} d_N(x^N, \mathbf{w}) > D + \delta/2} d\mathbf{w} f_{\mathbf{w}}(\mathbf{w}) d_N(x^N, \mathbf{w})]$$

$$\leq N \int dx^N f_{X^N}(x^N) [(D + \frac{\delta}{2}) + d_{\max} \int_{\mathbf{w}: N^{-1} d_N(x^N, \mathbf{w}) > D + \delta/2} d\mathbf{w} f_{\mathcal{W}}(\mathbf{w})]$$

$$= N[(D + \frac{\delta}{2}) + d_{\max} \int dx^N f_{X^N}(x^N) \int_{\mathbf{w}: N^{-1} d_N(x^N, \mathbf{w}) > D + \delta/2} d\mathbf{w} f_{\mathcal{W}}(\mathbf{w})],$$

where we have used the fact that the maximum distortion is given by $d_{\max}$. Define the probability

$$P(N^{-1} d_N(x^N, \mathcal{W}) > D + \delta/2) = \int_{\mathbf{w}: N^{-1} d_N(x^N, \mathbf{w}) > D + \delta/2} d\mathbf{w} f_{\mathcal{W}}(\mathbf{w}), \tag{3.4.6}$$

the probability that for a fixed vector x^N (selected by the source) there is no codeword in the code $\mathcal{W}$ (selected at random and independently of the source) that is within $D + \delta/2$ of the source word. Summarize the above bounds by

$$N^{-1} \bar{\rho}_N \leq D + \delta/2 + d_{\max} \int dx^N f_{X^N}(x^N) P(N^{-1} d_N(x^N, \mathcal{W}) > D + \delta/2). \tag{3.4.7}$$

The remainder of the proof is devoted to proving that the integral above goes to 0 as $N \to \infty$.

We begin with the probability $P(d_N(x^N, \mathcal{W}) > D + \delta/2)$ that for a fixed input block x^N, a randomly selected code will result in a minimum distortion codeword larger than $D + \delta/2$. This is the probability that none of the M words selected independently at random according to the pdf f_{Y^N} lie within $D+\delta/2$ of the fixed input word x^N:

$$P(N^{-1}d_N(x^N, \mathcal{W}) > D+\delta/2) = [1 - P(N^{-1}d_N(x^N, Z^N) \leq D+\delta/2)]^M \tag{3.4.8}$$

where Z^N is a random vector chosen according to the distribution f_{Y^N} for Y^N and

$$P(N^{-1}d_N(x^N, Z^N) \leq D+\delta/2) = \int_{y^N: N^{-1}d_N(x^N,y^N) \leq D+\delta/2} dy^N f_{Y^N}(y^N).$$

Unlike Y^N, Z^N is chosen independently of X^N and hence the above probabilities are not conditional on the value of X^N.

Now mutual information comes into the picture. The above probability can be bound below by adding a condition:

$$P(N^{-1}d_N(x^N, Z^N) \leq D+\delta/2)$$

$$\geq P(\frac{1}{N}d_N(x^N, Z^N) \leq D+\delta/2 \text{ and } \frac{1}{N}i_N(x^N, Z^N) \leq R-\epsilon/2),$$

where the sample information is computed according to the joint distribution for X^N, Y^N, that is,

$$i_N(x^N, z^N) = \log \frac{f_{X^N,Y^N}(x^N, z^N)}{f_{X^N}(x^N) f_{Y^N}(z^N)}. \tag{3.4.9}$$

Observe that there is a "mismatch" in that the sample information between X^N and the random vector Y^N is being evaluated at x^N and a random vector Z^N, the latter having the same marginal distribution as Y^N but not the same joint distribution with X^N as Y^N.

We now have the bound

$$P(\frac{1}{N}d_N(x^N, Z^N) \leq D+\delta/2)$$

$$\geq \int_{y^N: \frac{1}{N}d_N(x^N,y^N) \leq D+\delta/2 \text{ and } \frac{1}{N}i_N(x^N,y^N) \leq R-\epsilon/2} dy^N f_{Y^N}(y^N).$$

The information condition can be used to lower bound the pdf f_{Y^N} as

$$f_{Y^N}(y^N) \geq 2^{-N(R-\epsilon/2)} f_{Y^N|X^N}(y^N|x^N)$$

which yields the bound

$$P(\frac{1}{N}d_N(x^N, Z^N) \leq D+\frac{\delta}{2}) \geq$$

$$2^{-N(R-\epsilon/2)} \int_{y^N: N^{-1}d_N(x^N,y^N)\leq D+\frac{\delta}{2}, \frac{1}{N}i_N(x^N;y^N)\leq R-\frac{\epsilon}{2}} dy^N f_{Y^N|X^N}(y^N|x^N)$$

which with (3.4.8) gives

$$P(\frac{1}{N}d_N(x^N,\mathcal{W}) > D+\delta/2) = \left(1 - P(\frac{1}{N}d_N(x^N,Z^N) \leq D+\delta/2)\right)^M$$

$$\leq [1 - 2^{-N(R-\epsilon/2)} \times$$

$$\int_{y^N: \frac{1}{N}d_N(x^N,y^N)\leq D+\delta/2, \frac{1}{N}i_N(x^N,y^N)\leq R-\epsilon/2} dy^N f_{Y^N|X^N}(y^N|x^N)]^M$$

Applying the inequality

$$(1-\alpha\beta)^M \leq 1 + e^{-M\alpha} - \beta \quad (3.4.10)$$

for $\alpha, \beta \in [0,1]$ yields

$$P(\frac{1}{N}d_N(x^N,\mathcal{W}) > D+\delta/2) \leq$$

$$1 + \exp[-M2^{-N(R-\epsilon/2)}]$$

$$- \int_{y^N: N^{-1}d_N(x^N,y^N)\leq D+\delta/2, \frac{1}{N}i_N(x^N,y^N)\leq R-\epsilon/2} dy^N f_{Y^N|X^N}(y^N|x^N)$$

Recalling that $M = 2^{N(R-\epsilon/4)}$ results in

$$P(N^{-1}d_N(x^N,\mathcal{W}) > D+\delta/2) \leq 1 + \exp[-2^{N\epsilon/4}]$$

$$- \int_{y^N: N^{-1}d_N(x^N,y^N)\leq D+\delta/2, \frac{1}{N}i_N(x^N,y^N)\leq R-\epsilon/2} dy^N f_{Y^N|X^N}(y^N|x^N).$$

Observe that by construction the exponential term goes to 0 as $N \to \infty$. The remaining term is

$$\int_{y^N: N^{-1}d_N(x^N,y^N)> D+\delta/2 \text{ or } \frac{1}{N}i_N(x^N,y^N)> R-\epsilon/2} dy^N f_{Y^N|X^N}(y^N|x^N).$$

Summarizing to this point we now have that

$$\lim_{N\to\infty} \frac{1}{N}\bar{\rho}_N \leq D+\frac{\delta}{2}+d_{\max} \lim_{N\to\infty} \int dx^N f_{X^N}(x^N) P(N^{-1}d_N(x^N,\mathcal{W}) > D+\frac{\delta}{2})$$

$$\leq D+\frac{\delta}{2}+d_{\max} \lim_{N\to\infty} \int dx^N f_{X^N}(x^N) \times$$

$$\int_{y^N: N^{-1}d_N(x^N,y^N)>D+\delta/2 \text{ or } \frac{1}{N}i_N(x^N,y^N)>R-\epsilon/2} dy^N f_{Y^N|X^N}(y^N|x^N)$$

$$= D+\frac{\delta}{2}+d_{\max} \lim_{N\to\infty} P(\frac{1}{N}d_N(X^N,Y^N) > D+\frac{\delta}{2} \text{ or } \frac{1}{N}i_N(X^N,Y^N) > R-\frac{\epsilon}{2})$$

$$\leq D+\frac{\delta}{2}+d_{\max} \lim_{N\to\infty} P(\frac{1}{N}d_N(X^N,Y^N) > D+\frac{\delta}{2})+$$

$$d_{\max} \lim_{N\to\infty} P(\frac{1}{N}i_N(X^N,Y^N) > R-\frac{\epsilon}{2}),$$

where we have used the union bound $P(F \cup G) \leq P(F) + P(G)$ for the last inequality. Since the pdf for (X^N, Y^N) are drawn from a stationary and ergodic distribution, we have from the ergodic theorems for $N^{-1}d_N$ and $N^{-1}i_N$ that $N^{-1}d_N$ converges to $Ed_1(X_0,Y_0)$ and $N^{-1}i_N(X^N,Y^N)$ converges to $\bar{I}(X;Y)$ and hence from (3.4.4)–(3.4.5) the above limiting probabilities are 0. Thus we can choose N large enough to ensure that

$$\frac{1}{N}\bar{d}_N \leq D+\delta = D(R-\epsilon)+\delta,$$

which as previously argued implies there must exist a block code with nearly the same performance.

3.5 Synchronizing Block Codes

One potential problem with block codes is that one must know where the blocks begin in order to properly decode them. We point out two means of synchronizing the codes so that the decoder can locate the beginnings of the blocks. Synchronizing a block code also plays a role in the coding theorem for sliding-block codes to be considered in the next section.

One way to synchronize a block code of length, say, N, through a noiseless channel is to use a short *synchronization sequence* or *synch sequence* or *synch word* as a prefix to each block. The synch sequence should be very short relative to the block length so as not to waste too many bits and it should not appear anywhere in any of the remaining portions of the channel vectors. This restriction will reduce the rate of the code from R, but the reduction will be small if properly done. Say that we have a channel with transmission rate R. Construct a good block code of length N and rate $R-\epsilon$ for small $\epsilon > 0$. Since the DRF is continuous in R such a code exists with performance fairly close to $D(R)$. Since the rate is slightly smaller than R, not all the channel vectors must be used, only $2^{N(R-\epsilon)}$ and not all 2^{NR}. The channel vectors have length K where

$$R = \frac{K}{N}\log||B||,$$

which we assume to be an integer. Suppose that we only use those channel vectors of length K for which the first $K\delta$ symbols form a special synchronization or synch sequence and the remaining $K(1-\delta)$ symbols nowhere contain the synch word. We also must ensure that no overlap of part of the synch word with a code word can appear as a synch word. If this restriction is met, the decoder can always locate the block beginnings and decode correctly by simply finding a synch sequence. (In fact, once the synch sequence is found, the blocks are in theory all synchronized.) The question is then how many such words are there? In particular, are there enough to provide indices for all $2^{N(R-\epsilon)}$ codewords?

To show that enough words remain to communicate the code, choose as a particular example a synch word consisting of $K\delta$ repetitions of a given channel symbol, say 0. To ensure that the first and last symbol of the remaining $K(1-\delta)$ symbols in the channel block cannot be confused as part of a synch word, force these to be another channel symbol, say 1. We now have $K(1-\delta)-2$ free channel symbols to use subject to the constraint that there not appear $K\delta$ consecutive 0's in any channel block of this length. There are a total of $||B||^{K(1-\delta)-2}$ channel K-tuples having the synch sequence as a prefix and having a 1 at the beginning and end of the remaining block. Of these, no more than $(K(1-2\delta)-1)||B||^{K(1-2\delta)-2} \leq K||B||^{K(1-2\delta)-2}$ have the synch sequence appearing somewhere in the vector (there are $K(1-2\delta)-1$ possible locations for a complete synch sequence of $K\delta$ symbols to start in a block of $K(1-\delta)-2$ symbols and the remaining $K(1-2\delta)-2$ symbols are arbitrary). Thus the total number of channel vectors available is at least

$$||B||^{K(1-\delta)-2} - K||B||^{K(1-2\delta)-2} = ||B||^{K(1-\delta)-2}(1-K||B||^{-\delta K}),$$

and hence the available rate using the constrained words is

$$\begin{aligned} R' &= \frac{K(1-\delta)-2}{N}\log||B|| + \frac{1}{N}\log(1-K||B||^{-\delta K}) \\ &= R(1-\delta) - \frac{2\log||B||}{N} + \frac{1}{N}\log(1-K||B||^{-\delta K}). \end{aligned}$$

Given R and ϵ as before, choose $\delta > 0$ and then N and hence K large enough to ensure that $R' > R-\epsilon$ (while preserving the relation that $R = (K/N)\log||B||$). This proves that the sychronized codebook has sufficient rate to include all of the words in the unsynchronized codebook.

Thus one way to achieve synchronization is to use a short synchronization sequence as a prefix to all channel code words.

Another way to synchronize is to use a longer synch sequence, but use it very rarely, e.g., to add a synch sequence to the front of every group of

L channel code vectors. Here, for example, given a block code of length N and channel word length K, we might select a synch word consisting of $3K$ repetitions of a single channel symbol b. If the available channel vectors are pruned so that none begin or end with this symbol (analogous to the more complicated pruning above), then the decoder can locate the block beginnings by finding the synch sequences. Again the rate is reduced slightly, but now also some source symbols will be lost while the synch sequence is being sent and hence the distortion is also increased. If the synch sequence is sent rarely enough, the loss will be slight.

3.6 Sliding-Block Codes

We next use the block coding results to sketch a similar result for sliding-block codes. For simplicity we consider the case where the time duration of source and channel symbols is the same ($\tau_s = \tau_c$) and the channel alphabet corresponds to an integral number of bits, that is, $||B|| = 2^R$ for an integer R. Thus in this case $R_{\text{trans}} = R$ bits per source symbol. We will later indicate how this restriction is removed. In this special case a sliding-block encoder looks at a window of source symbols to produce a binary R-tuple for the channel. It then slides the window over by one source symbol and produces the next channel vector. The decoder is assumed to be a similar sliding-block code, except that it windows a group of channel "symbols," each of which is a binary R-tuple, to produce a single reproduction symbol. It then slides the window by one symbol, i.e., one binary R-tuple, and produces the next reproduction symbol. Thus both encoder and decoder can be viewed as shift registers with a nonlinear mapping producing the output. We will argue that there exist such codes which perform at least as well as block codes and hence arbitrarily near the DRF. Again the cost is that we may need large window lengths and complicated nonlinear mappings. Observe that such a code will yield a stationary source/reproduction process $\{X_n, \hat{X}_n\}$ and that, in particular, the decoder consists of a binary shift register and a nonlinear mapping producing the reproduction. Each successive reproduction is produced by shifting the contents of the shift register R times.

While the details of the proof are involved, the basic idea is straightforward: Form a good synchronized block code of length, say, N, for the source. This code has rate R and, if used as an ordinary block code, has average distortion $N^{-1}\bar{d}_N$ nearly $D(R)$. The block code is used to construct a sliding-block or stationary code by occasionally inserting random spacing between the blocks. To accomplish this, select an M-dimensional source event, that is, a collection of vectors x^M with the property that the probability of this event is very small but not 0. First construct a sliding-block code with a dependence on the infinite past as follows: given the current symbol, say X_n, look at vectors $X_{n-M-i+1}, \ldots, X_{n-i}$ for $i = 0, 1, \ldots$ until a synch event corresponding to the occurrence of the synch sequence in the input stream is seen. If the current symbol is not in that synch pattern, use the synchronized block code to code from the pattern up to and including the current symbol. In other words, begin parsing the input sequence into blocks at the end of the synch sequence and continue to code successive blocks until the block containing the current symbol is coded. The channel symbol sent will be that corresponding to the current position in the block code word. If the current symbol is in a synch pattern, send an arbitrary "filler" channel letter. The encoder will send sequences of block code words that can be decoded by the decoder (since they are synchronized) along with occasional incomplete blocks and filler symbols which make the code stationary. The distortion will decrease slightly because of the symbols lost during the synch pattern or during incomplete blocks interrupted by a synch pattern.

This code has infinite memory because there is no bound to how far back one might have to look to find a synch pattern. The performance of the code can be approximated, however, by a finite memory code that looks sufficiently far back and gives up if it does not find a pattern. It can be shown that the distortion increases only slightly because only rarely will no synch pattern be seen in the window.

The above is certainly not a practicable scheme, but it does demonstrate that good sliding-block codes exist. Unlike block codes, however, there do not exist general algorithms for designing good sliding block codes and that most such codes have been developed from heuristic arguments.

3.7 Trellis Encoding

Suppose now that we have a good sliding-block code, that is, one with transmission rate R and performance near the distortion-rate function. Focus now on the decoder, which can be viewed as a shift register which holds, say, K channel symbols together with a nonlinear mapping f of B^K into a source symbol or group of source symbols. For simplicity we consider the case of a binary channel with transmission rate 1 bit per source symbol. Thus at each unit of time R channel symbols are mapped into a single source symbol, a fact which we can write as

$$\hat{X}_n = f(U_{n-M}, \ldots, U_n, \ldots, U_{n+D}),$$

where M is the *memory length* of the decoder and D is the *delay*. For convenience we reindex the channel symbols so that the nonlinear filter described by f is causal, that is, we write

$$\hat{X}_n = f(U_{n-l}, \ldots, U_n),$$

where l is $M + D$. Alternatively, the decoder can be viewed as a finite state machine, where the state of the shift register corresponds to all but the newest of the channel symbols in the register; that is, we define the state S_n of the decoder at time n as the vector $U_{n-l}, \ldots, U_{n-1}$. Then we have a finite state code representation with output

$$\hat{X}_n = f(U_n, S_n)$$

and the next-state mapping is given by $g(a, s_n) = s = (u_1, u_2, \ldots, u_{K-1}, a)$ if $s_n = u_0, u_1, \ldots, u_{K-1}$; that is, one channel symbol is shifted out and the new symbol shifted in.

The action of the decoder can be depicted by a directed graph called a tree as in Figure 3.1, where the channel is assumed to have a binary alphabet (binary tree) the shift register length is 3 and hence there are $2^{3-1} = 4$ states. The outputs can then be thought of as functions of the state (which is 00, 01, 10, or 11) and the encoded symbol (which is 1 or 0). Thus at each node the tree can advance to one of two available nodes (because there are two possibilities for the code symbols), but the actual outputs depend on binary triples.

Suppose that the decoder starts in the state where the shift register contains only zeroes. A binary channel symbol forces a state change and produces an output, the choices being shown on the first two branches of the tree. Moving up in the tree corresponds to a channel symbol of 1 and moving down corresponds to a 0. The nodes in the diagram correspond to the states of the decoder and the labels on the branches correspond to the decoder outputs when making the corresponding state transitions.

Now forget the original sliding-block encoder and consider the following alternative: If the encoder has a copy of the decoder, it could take an input sequence of, say, length L and try to find the minimum distortion path through the decoder tree to a depth of L, that is, to compare all of the possible paths through the tree and find the one which produces a sequence of labels that has the smallest distortion with respect to the input sequence. The corresponding channel symbols (which can be thought of as a path map describing how to traverse the tree) are then sent to the decoder, which produces the reproduction sequence. In other words, the encoder finds a binary sequence that will produce a sequence of L reproduction symbols which best match the input source L symbols. It then drives the decoder

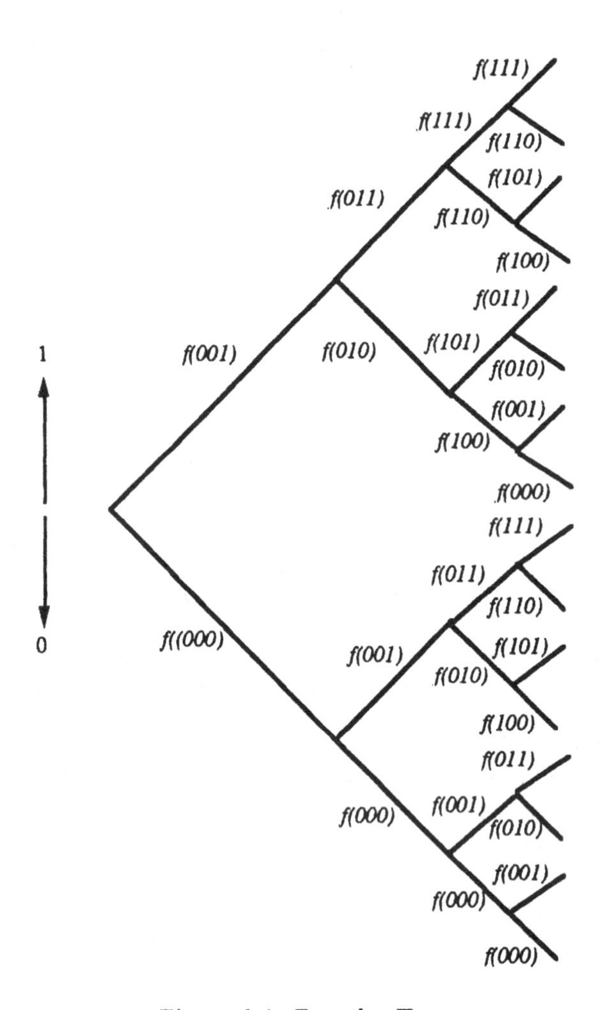

Figure 3.1: Decoder Tree

with the sequence. The decoder can then be reset to an original state and the process repeated. Note that this new encoder is effectively operating as a block code of length L. This problem, finding a minimum cost path through a tree, has been much studied in a variety of applications and such algorithms provide encoders matched to a particular decoder. Before considering examples of such algorithms, we consider the implication for coding theorems. If we know that a sliding-block encoder yields an long term average distortion of about $D(R)$ using the given decoder and we then replace that encoder by a tree search encoder which finds the minimum distortion path, one would expect that for sufficiently large L the tree search encoder should perform no worse than the original encoder since it is finding the best possible path. This idea can be made rigorous to prove that there exist good tree encoding systems, but the details are complicated by the fact that the original sliding-block code yields a stationary process while the tree search code used as above does not.

In the case of a finite state decoder, the tree diagram is highly redundant, a fact which permits a simplification. In particular, all possible decoder states have appeared by the $(K-1)$st level of the tree and hence many nodes in each level thereafter represent the same state. If we merge these identical nodes we obtain a special form of tree called a *trellis* as shown in Figure 3.2.

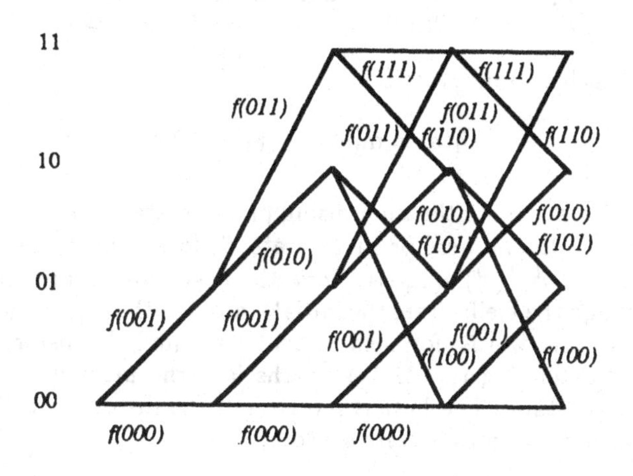

Figure 3.2: Trellis

The trellis provides a useful picture of the time evolution of the decoder: each node corresponds to a decoder state and each column to a time instant. Since the decoder definition does not change with time, the trellis is also redundant in that a branch connecting two given states always has the same label regardless of the time. Denote by $g(s', s)$ the label of the branch connecting an old state s' to a new state s and denote the channel symbol corresponding to this transition $i(s', s)$.

The best path through a trellis can be found using the Viterbi algorithm which is an application of the principle of optimality of dynamic programming. (See Forney [27] for a general development of the Viterbi algorithm. Forney brought the word *trellis* from gardening to coding theory by naming the merged tree a trellis and trellis coding systems for sources and channels owe much to him.) In our application the key idea is this: The minimum distortion path from time 0 to time n must be an extension of one of the minimum distortion paths to a node at time $n-1$. Thus in order to find the best possible path of length L we compute the best path into each state for each time unit by finding the best extension from the previous states into the current state and we perform this for each time unit up until time L. We formalize this procedure as an algorithm:

Viterbi algorithm trellis encoder

1. Given a collection of states $\mathcal{S} = \sigma_0, \ldots, \sigma_M$, a starting state σ', a decoder $f(u, s)$ producing a reproduction given channel symbol u in state s, a per letter distortion measure d, a source input vector $x_0, \ldots, x_{L-1}$. Let $D_j(k)$ denote the distortion for state k at time j. Set $D(\sigma', 0) = 0$, $D(s, 0) = \infty$ for $s \neq \sigma'$. Set $l = 1$.

2. For each current state s find

$$D_l(s) = \min_{s'} \left(D_{l-1}(s') + d(x_l, g(s', s)) \right)$$

   and let s' denote the minimizing previous state. Given the path map $u^{l-1}(s')$ to the best previous state s', form the extended path map $u^l(s) = (u^{l-1}(s'), i(s', s))$ giving the best (minimum distortion) path through the trellis from the initial state into the current state s at level l. In other words, for each current state find the distortion resulting from extending the M best paths into the previous states into the current state. The best is picked and the distortion and cumulative path indices to that state saved.

3. If $l < L$, set $l + 1 \rightarrow l$ and go to 1. If $l = L$, pick the final state s_f yielding the minimum of the M values $D_L(s)$. Release $u^L(s_f)$.

A key facet above is that we have a block code with block length L, but we do not have to search a block code of 2^L by computing all 2^L

possible distortions. Instead we compute a sequence of M distortions and retain the running distortion for each state along with the accumulated distortions into each state and the accumulated channel symbol sequence. In particular, the complexity grows with the number of states, not with L.

Trellis encoders are also known as *look ahead coders* since they permit the source to view the effects of possible channel sequences into the future. Other names are *multipath search encoders* and *delayed decision encoders*. They provide a reasonable complexity means of having large blocklength codes, but that also has the effect of causing large decoding delays which can be harmful in some cases, e.g., in control loops.

If the sliding-block decoder has infinite memory, then the decoder is not a finite state machine and the Viterbi algorithm cannot be used. In this case, other tree search algorithms can be used. In particular, the (M, L)-algorithm or M-algorithm of Jelinek and Anderson [53] provides a suboptimal (it does not always find the best path, just a good one) but efficient means of searching a tree or trellis.

Exercises

1. The Nth order *rate-distortion function* of a source $\{X_n\}$ and distortion measure d_N is defined by

$$R_N(D) = \inf I(X^N; \hat{X}^N)$$

where the infimum is over all conditional probability distributions for $\hat{X}^N$ given X^N for which

$$\frac{1}{N} E d_N(X^N, \hat{X}^N) \leq D.$$

(a) For fixed $R > 0$, relate $R_N(D_N(R))$ to R and for a fixed D relate $D_N(R_N(D))$ to D.

(b) Suppose that $\{X_n\}$ is an iid binary process with $p_X(1) = 1 - p_X(0) = p$, and that d_1 is a Hamming distortion measure. Let $+$ denote modulo two addition ($0+0 = 1+1 = 0$, $0+1 = 1 + 0 = 1$).

(c) Show that

$$H(X_0|Y_0) = H(X_0 + Y_0|Y_0) \leq H(X_0 + Y_0),$$

(d) Show that

$$R_1(D) \geq H(X_0) - \max_{p_{Y_0|X_0}} H(X_0|Y_0)$$

where the maximum is over all conditional pmf's $p_{Y_0|X_0}$ with $Ed_1(X_0, Y_0) \leq D$.

(e) Show that

$$R_1(D) \geq H(X_0) - \max H(Z)$$

where the maximum is over all binary pmf's p_Z for which $EZ \leq D$.

(f) Find a simple lower bound to $R_1(D)$ in terms of p and D. What does this imply about $D_1(R)$?

(g) Find a simple lower bound to $R_N(D)$ from the above bound to $R_1(D)$.

2. Use a continuous analogy to the argument of the above problem to derive a lower bound to the first order rate-distortion and distortion-rate functions for an iid Gaussian random process with 0 mean and variance σ^2. (*Hint:* Use problem 2.5.)

3. Prove (3.4.10).

4. A binary random process $\{X_n\}$ has the following property: For a fixed large integer N it is known that any time a 1 occurs, it will necessarily be followed by at least N 0's. Thus the process output tends to produce mostly 0's with occasional isolated ones. Find bounds to the first order entropy $H(X_0)$ and the entropy rate $\bar{H}(X)$ of the process. Find an almost noiseless block code with a rate or required transmission rate that is close to the optimal value. How could you use this process to construct a sliding-block code from a block code?

5. Prove (3.4.10).

6. In this problem it is shown that for an iid process X_n

$$D(R) = \frac{D_N(R)}{N} = D_1(R)$$

for all positive integers N. Thus the DRF can be evaluated as a one dimensional minimization.

(a) Suppose first that the conditional pmf $p_{Y_0|X_0}$ achieves the minimization defining $D_1(R)$, that is,

$$\sum_{x,y} d_1(x, y) p_{Y_0|X_0}(y|x) p_{X_0}(x) = D_1(R)$$

and

$$\sum_{x,y} p_{Y_0|X_0}(y|x) p_{X_0}(x) \log \frac{p_{Y_0,X_0}(y,x)}{p_{X_0}(x) p_{Y_0}(y)} \leq R.$$

(b) Define the N-dimensional conditional pmf $p_{Y^N|X^N}$ by the product pmf

$$p_{Y^N|X^N}(y^N|x^N) = \prod_{k=0}^{N-1} p_{Y_0|X_0}(y_k|x_k).$$

Show that this pmf yields a mutual information

$$\frac{1}{N} I(X^N; Y^N) \leq R$$

and an average distortion

$$Ed_N(X^N, Y^N) = N D_1(R).$$

Show that this implies that

$$D_1(R) \geq \frac{D_N(R)}{N}.$$

(c) Conversely, suppose that $p_{Y^N|X^N}$ yields $D_N(R)$, that is,

$$\frac{1}{N} I(X^N; Y^N) \leq R$$

and

$$Ed_N(X^N, Y^N) = D_N(R).$$

Define

$$R_k = I(X_k; Y_k)$$

and show that

$$Ed_N(X^N, Y^N) \geq \sum_{k=0}^{N-1} D_1(R_k)$$

and

$$\frac{1}{N} \sum_{k=0}^{N-1} R_k \leq R.$$

Use the convexity of DRFs to then argue that

$$N^{-1} D_N(R) \geq D_1(R),$$

which completes the proof.

7. Use the previous problem to prove the claim of (3.4.1)–(3.4.2) for the case of a discrete alphabet iid source.

Chapter 4

Rate-Distortion Functions

4.1 Basic Properties

We have seen that the distortion-rate function of a source with respect to a distortion measure provides an unbeatable, yet approximately achievable, bound on the performance of a fixed-rate data compression system with unconstrained block length or delay. The goal of this chapter is to provide the tools for actually computing and bounding the distortion rate functions for particular sources and distortion measures. Toward this end we begin by developing variational equations for inverse distortion-rate functions or rate-distortion functions. The focus on rate-distortion functions instead of distortion-rate functions is both for consistency with the literature and for simplicity. It is usually easier to find the rate as a function of distortion rather than vice-versa.

Explicit evaluations of the rate-distortion function are often not possible, even for apparently simple sources and distortion measures. Hence we also develop a variety of simple bounds to rate-distortion functions and a numerical algorithm for evaluating finite order rate-distortion functions.

In the first three sections we concentrate on the first-order rate-distortion function of a discrete alphabet random process, that is, on the rate-distortion function of a single random variable. The results will generalize to higher order rate-distortion functions by application of the same techniques to vectors. The extensions to continuous alphabet processes will be sketched and the results quoted. The continuous results are usually exact parallels to the discrete results with pmfs and sums being replaced by the corresponding pdfs and integrals. The emphasis on discrete alphabets reflects the fact that numerical algorithms for the computation of rate-distortion functions are based in part on the variational equations and such computation is likely to be accomplished on a digital computer and hence on a discrete approximation to a continuous distribution.

Given a discrete random variable X, abbreviate by p_j the pmf $p_X(j) = \Pr(X = j)$ for $j \in A$, the source alphabet. Given a reproduction alphabet $\hat{A}$ and a conditional probability mass function $Q_{k|j} = \Pr(Y = k|X = j)$, define the average mutual information

$$I(p, Q) = \sum_{j,k} p_j Q_{k|j} \log \frac{Q_{k|j}}{\sum_i p_i Q_{k|i}}, \tag{4.1.1}$$

where the notation emphasizes the dependence on the input and conditional distributions. We will usually abbreviate this to simply $I(Q)$ since we consider p to be fixed while we optimize over Q. Let d be the distortion measure on $A \times \hat{A}$. Let $\mathcal{Q}_{\mathcal{D}}$ denote the collection of all valid conditional pmf's Q (i.e., $\{Q_{k|j},\ j \in A, k \in \hat{A}\}$ such that $Q_{k|j}$ is nonnegative and for any fixed j the sum over k is 1) which satisfy the average distortion constraint

$$\bar{d}(p, Q) = \sum_{k,j} Q_{k|j} p_j d(j,k) \leq D. \tag{4.1.2}$$

We similarly abbreviate this to $\bar{d}(Q)$. Then the rate-distortion function is defined by

$$R(D) = \min_{Q \in \mathcal{Q}_{\mathcal{D}}} I(Q). \tag{4.1.3}$$

We assume for convenience that

$$\min_k d(j,k) = 0;$$

that is, the minimum possible distortion is 0. It can be shown that this entails no real loss of generality since results under this assumption can easily be extended to more general distortion measures by defining a modified distortion measure $d'(j,k) = d(j,k) - \min_k d(j,k)$, which satisfies the assumption of 0 minimum distortion. Given this assumption, the minimum possible average distortion will also be 0. To see this, for each j set $Q_{k|j} = 1$ for the k yielding the minimum distortion.

As D increases from 0, $\mathcal{Q}_{\mathcal{D}}$ is a nondecreasing collection and hence $R(D)$ must be monotonically nonincreasing in D. D will increase up to some value D_{max} which is defined as the smallest distortion that can be achieved when $R(D) = 0$. $I(p, Q)$ will be 0 if and only if the input and output are independent, in which case $Q_{k|j} = Q_k$ and

$$\bar{d}(p, Q) = \sum_j \sum_k p_j Q_k d(j,k).$$

This expression is minimized over all valid pmf's Q by setting $Q_k = 1$ for the k which minimizes $\sum_j p_j d(j,k)$ and setting the remaining $Q_k = 0$. This yields

$$D_{\max} = \min_k \sum_j p_j d(j,k). \tag{4.1.4}$$

Note that this rate-distortion pair is achieved by the 0 rate code consisting of one symbol, the k which minimizes the average distortion $\sum_j p_j d(j,k)$ as above. For all values of distortion D larger than $D_{\max}$ the same conditional pmf yields both a distortion smaller than D and a mutual information of 0. Thus

$$R(D) = 0, D \geq D_{\max}.$$

There is another, more operational, way to view $D_{\max}$. Suppose that we wish to construct the best possible code with 0 rate, that is, we are permitted only one code word, say y, and we want to choose it to minimize $Ed(X,y)$. Choosing y to achieve this minimum results in an average distortion of exactly $D_{\max}$.

A key property of the rate-distortion function is its convexity, which is described in the following lemma.

Lemma 4.1.1 *$R(D)$ is a convex function of D, that is, for any two values of distortion D_1 and D_2 and any $\lambda \in (0,1)$*

$$R(\lambda D_1 + (1-\lambda)D_2) \leq \lambda R(D_1) + (1-\lambda)R(D_2).$$

Proof: Suppose that Q^1 and Q^2 achieve the points $(D_1, R(D_1))$ and $(D_2, R(D_2))$ on the rate-distortion curve, respectively. Define the conditional pmf $\bar{Q} = \lambda Q^1 + (1-\lambda)Q^2$. By the linearity of expectation

$$\bar{d}(\bar{Q}) = \lambda \bar{d}(Q^1) + (1-\lambda)\bar{d}(Q^2)$$

whence $\bar{Q} \in \mathcal{Q}_{\lambda D_1 + (1-\lambda)D_2}$. Thus

$$R(\lambda D_1 + (1-\lambda)D_2) \leq I(\bar{Q}).$$

It can be shown using the divergence inequality that $I(Q)$ is convex in Q and hence that

$$I(\bar{Q}) \leq \lambda I(Q^1) + (1-\lambda)I(Q^2) = \lambda R(D_1) + (1-\lambda)R(D_2),$$

which completes the proof.

Since $R(D)$ is convex, it is also continuous, except possibly at the endpoint $D = 0$. Since it is convex, $R(D)$ must be strictly decreasing and not just nonincreasing for $D \in (0, D_{\max})$. This implies further that if Q yields

$R(D)$, then $\bar{d}(Q) = D$ and hence the optimizing Q in the collection $\mathcal{Q}_D$ of all Q such that $\bar{d}(Q) \leq D$ is on the boundary of that collection. This is true since if $\bar{d}(Q)$ were equal to some $D' < D$, then $R(D) = I(Q) \geq R(D')$, which contradicts the fact that $R(D)$ is decreasing in D. This provides the relation between rate-distortion and distortion-rate functions as summarized in the following lemma:

Lemma 4.1.2 *The rate-distortion function and the distortion-rate function are inverses, that is, if $R(\delta) = r$, then $D(r) = \delta$. Alternatively, $R(D(r)) = r$ and $D(R(\delta)) = \delta$.*

Proof: Suppose that Q achieves $r = R(\delta)$ and hence a point (δ, r) on the rate-distortion curve. Since $I(Q) = r$ and $\bar{d}(Q) = \delta$, we must have that $\delta \geq D(r)$. Suppose that the inequality is strict. In this case there is a Q' with $I(Q') \leq r$ and $\bar{d}(Q') = D(r) < \delta$ and therefore

$$R(\delta) = r \geq I(Q') \geq R(D(r)).$$

This, however, violates the fact that $R(D)$ is strictly decreasing in D, hence $\delta = D(r)$.

As a final property, suppose that $A \subset \hat{A}$, that is, the reproduction alphabet contains a copy of the source alphabet and suppose further that $d(j,k)$ is 0 if and only if $k = j$. Then the only Q yielding $\bar{d}(Q) = 0$ is that which sets $Q_{j|j} = 1$ for each j and hence yields $R(0) = H(p)$, the source entropy.

In summary, we have shown that the rate-distortion function (and hence also the distortion-rate function) are monotonically decreasing convex functions of their argument and hence are also continuous functions except possibly at the endpoints. These properties are summarized pictorially in a typical $R(D)$ curve in Figure 4.1.

4.2 The Variational Equations

Shannon provided the first evaluations of rate-distortion functions for simple memoryless sources and difference distortion measures [74], but the general theory of evaluating rate-distortion functions was first developed by Gallager [28], Chapter 9, and expanded by Berger [5]. The development in this chapter largely follows Blahut [6], but the proofs vary somewhat from his in order to take more advantage of the divergence inequality (Lemma 2.6.1) and to avoid minimizations by calculus.

The minimization of (4.1.3) can be accomplished by introducing a Lagrange multiplier s and replacing the minimization of (4.1.3) by parametric expressions for the rate and distortion. These parametric expressions

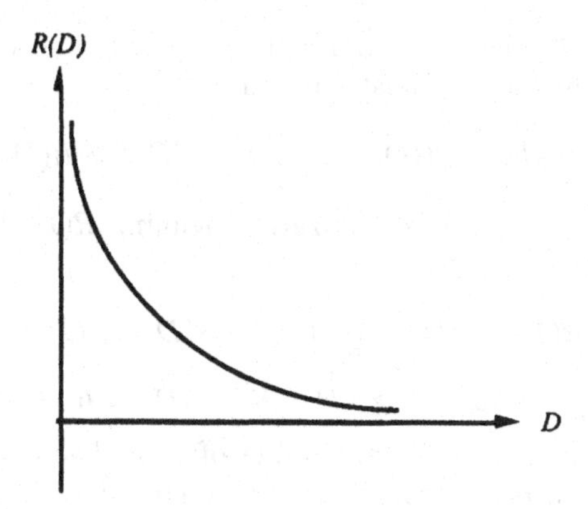

Figure 4.1: Typical Rate-Distortion Function

in turn yield the rate-distortion function. The following theorem gathers these pieces. Parts (a) and (c) are proved, but (b) is not.

Theorem 4.2.1 *For each $s \leq 0$, define a rate-distortion pair (R_s, D_s) parametrically by*

$$R_s = sD_s + \min_Q \left[I(Q) - s\bar{d}(Q)\right] \tag{4.2.1}$$

$$D_s = \bar{d}(Q) = \sum_j \sum_k p_j Q^*_{k|j} d(j,k), \tag{4.2.2}$$

where Q^* is the conditional pmf yielding the minimum in (4.2.1).

a *Each value of the parameter s leads to a rate-distortion pair (R_s, D_s) on the rate-distortion curve, that is,*

$$R_s = R(D_s) \tag{4.2.3}$$

(or $D_s = D(R_s)$).

b *Every point on the rate-distortion curve has this form for some s, that is, for any $D \in [0, D_{\max}]$ there is an s for which (4.2.1)–(4.2.2) hold.*

c *The rate-distortion function can be found from the nonparametric relation*

$$R(D) = \sup_{s \leq 0}(R_s + s(D - D_s)).$$

Proof:

Suppose that Q^* yields the minimum of (4.2.1) and hence that $D_s = \bar{d}(Q^*)$. Then for this Q^* we must have that

$$R_s = sD_s + (I(Q^*) - sD_s) = I(Q^*) \geq R(D_s).$$

Conversely, if Q^* solves the minimization defining $R(D)$ then $\bar{d}(Q^*) \leq D$ and we have for any $s \leq 0$ that

$$R(D) = I(Q^*) \geq I(Q^*) + s(D - \bar{d}(Q^*))$$

$$= sD + I(Q^*) - s\bar{d}(Q^*) \geq sD + R_s - sD_s = R_s + s(D - D_s).$$

This expression is true for any value of D and any value of s. This provides two conclusions. First, given s and hence (R_s, D_s), choosing D as D_s yields

$$R(D_s) \geq R_s,$$

proving (4.2.3) and hence (a). Second, the bound still holds if we maximize it over s and hence

$$R(D) \geq \sup_{s \leq 0}(R_s + s(D - D_s)).$$

Part (c) follows from this inequality together with (b) which implies that the inequality must be an equality for some s.

A key facet of the variational description is that the minimization is over all conditional pmfs Q and not just those in $\mathcal{Q}_\mathcal{D}$. It can be shown that s is precisely the slope $R'(D)$ of the rate-distortion curve at D_s.

The following theorem, which is due to Blahut [6], provides the basic relations required by the numerical algorithm for computing rate-distortion (or distortion-rate) curves and also yields the basic variational equations for explicit evaluation.

Theorem 4.2.2 *Given the source pmf p and a conditional pmf Q, let r denote the induced output pmf*

$$r_k = \sum_j p_j Q_{k|j},$$

and let q denote an arbitrary pmf on the output alphabet $\hat{A}$ (not necessarily the same as r). Define the functional

$$F(p, Q, q) = \sum_j \sum_k p_j Q_{k|j} \log \frac{Q_{k|j}}{q_k} - s \sum_j \sum_k p_j Q_{k|j} d(j, k). \qquad (4.2.4)$$

Then

1. *the rate-distortion function is given parametrically as*

$$R_s = sD_s + \min_q \min_Q F(p, Q, q) \tag{4.2.5}$$

$$D_s = \sum_j \sum_k p_j Q^*_{k|j} d(j,k), \tag{4.2.6}$$

where Q^* is the minimizing conditional pmf.

2. *for fixed Q, $F(p,Q,q)$ is minimized by*

$$q_k = \sum_j p_j Q_{k|j}, \tag{4.2.7}$$

3. *for fixed q, $F(p,Q,q)$ is minimized by*

$$Q_{k|j} = \frac{q_k e^{sd(j,k)}}{\sum_l q_l e^{sd(j,l)}}. \tag{4.2.8}$$

A distribution of this form is sometimes called a tilted distribution.

Proof: (a) We have with some algebra that

$$F(p,Q,q) + s\sum_j \sum_k p_j Q_{k|j} d(j,k) - I(p,Q) = H(r||q)$$

and hence from the divergence inequality (Lemma 2.6.1)

$$F(p,Q,q) \geq I(p,Q) - s\sum_j \sum_k p_j Q_{k|j} d(j,k)$$

with equality if and only if $q = r$. This proves (a) since it implies that the stated minimization is equivalent to the previous parametric minimization for $R(D)$.

(b) This follows from the equality condition of part (a).

(c) Observe that we can rewrite $F(p,Q,q)$ as

$$F(p,Q,q) = \sum_j p_j \left(\sum_k Q_{k|j} \log \frac{Q_{k|j}}{q_k e^{sd(j,k)}} \right).$$

This resembles a relative entropy except that the potential pmf in the denominator of the logarithm may not sum to one. Thus if we define a new conditional pmf by

$$Q'_{k|j} = \frac{q_k e^{sd(j,k)}}{\sum_l q_l e^{sd(j,l)}},$$

then

$$F(p,Q,q)=\sum_j p_j H(Q_{\cdot|j}\|Q'_{\cdot|j})-\sum_j p_j\log\sum_l q_l e^{sd(j,l)}.$$

The rightmost term does not depend on Q. The relative entropies are all bound below by 0 and achieve that value if and only if Q is chosen to equal Q'. This proves (c). Note that in fact we need only have $Q_{\cdot|j}=Q'_{\cdot|j}$ for those j for which $p_j>0$ in order to have

$$F(p,Q,q)=-\sum_j p_j\log\sum_l q_l e^{sd(j,l)}$$

The theorem yields many corollaries describing $R(D)$. The first follows immediately from the final paragraph of the proof of the theorem.

Corollary 4.2.1 *Given the parameter s, the corresponding (R_s,D_s) point on the rate-distortion curve is given by*

$$R_s=sD_s+\min_q\Big(-\sum_j p_j\log\sum_l q_l e^{sd(j,l)}\Big) \tag{4.2.9}$$

$$D_s=\sum_j\sum_k p_j Q^*_{k|j}d(j,k) \tag{4.2.10}$$

where

$$Q^*_{k|j}=\frac{q^*_k e^{sd(j,k)}}{\sum_l q^*_l e^{sd(j,l)}}, \tag{4.2.11}$$

where q^* solves the minimization of (4.2.8).

Simultaneous satisfaction of both parts (b) and (c) of the theorem yield the following corollary characterizing the optimal conditional pmf's.

Corollary 4.2.2 *If Q is a conditional pmf which achieves the point on the $R(D)$ curve corresponding to the parameter s, then*

$$Q_{k|j}=\frac{q_k e^{sd(j,k)}}{\sum_l q_l e^{sd(j,l)}}, \tag{4.2.12}$$

where

$$q_k=\sum_j p_j Q_{k|j}=q_k\sum_j p_j\frac{e^{sd(j,k)}}{\sum_l q_l e^{sd(j,l)}}. \tag{4.2.13}$$

Thus a necessary condition for the minimum to be achieved is that

$$c_k\equiv\sum_j p_j\frac{e^{sd(j,k)}}{\sum_l q_l e^{sd(j,l)}}=1 \text{ if } q_k\neq 0. \tag{4.2.14}$$

The following lemma shows how (4.2.14) behaves when $q_k = 0$.

Lemma 4.2.1 *Given the conditions of the previous corollary,*

$$c_k \equiv \sum_j p_j \frac{e^{sd(j,k)}}{\sum_l q_l e^{sd(j,l)}} \leq 1 \textit{ all } k. \tag{4.2.15}$$

Proof:
From the previous corollary we need only consider the case where $q_k = 0$. Suppose that the minimum over q and Q of the previous theorem are achieved yielding

$$F_o = F(p, Q, q) = \sum_j p_j \log \frac{1}{\sum_l q_l e^{sd(j,l)}}$$

and that $q_k = 0$ and

$$c_k = \sum_j p_j \frac{e^{sd(j,k)}}{\sum_l q_l e^{sd(j,l)}} > 1.$$

We shall show that this leads to a contradiction. In particular, we shall show that F_o can be decreased slightly by making $q_k > 0$ and hence it cannot be the minimum as assumed. To accomplish this suppose that $\epsilon > 0$ is small (how small will be specified later) and define the pmf

$$t_l = \begin{cases} (1-\epsilon)q_l & \text{if } l \neq k \\ \epsilon & \text{otherwise} \end{cases},$$

that is, we increase q_k slightly away from 0 and we decrease all of the other q_l by a corresponding amount so that the sum over the new pmf is 1. This new pmf results in

$$F(p, Q, t) = \sum_j p_j \log \frac{1}{\sum_l t_l e^{sd(j,l)}} = -\sum_j p_j \log \left[(1-\epsilon) \sum_l q_l e^{sd(j,l)} + \epsilon e^{sd(j,k)} \right]$$

$$= -\sum_j p_j \log \left[\left(\sum_l q_l e^{sd(j,l)} \right) \left((1-\epsilon) + \epsilon \frac{e^{sd(j,k)}}{\sum_l q_l e^{sd(j,l)}} \right) \right]$$

$$= F_o - \sum_j p_j \log \left((1-\epsilon) + \epsilon \frac{e^{sd(j,k)}}{\sum_l q_l e^{sd(j,l)}} \right)$$

$$= F_o - \sum_j p_j \log \left(1 + \epsilon \left[\frac{e^{sd(j,k)}}{\sum_l q_l e^{sd(j,l)}} - 1 \right] \right).$$

We can use a Taylor series to expand $\log(1+\delta_j)$ for small δ_j as

$$\log(1+\delta_j) \approx \delta_j - \frac{1}{2}\delta_j^2 + \frac{1}{3}\delta_j^3 + \cdots$$

and hence choosing ϵ sufficiently small to guarantee that all of the second order terms are negligible we have that

$$F(p,Q,t) \approx F_o - \epsilon(\sum_j p_j[\frac{e^{sd(j,k)}}{\sum_l q_l e^{sd(j,l)}} - 1]) = F_o - \epsilon(c_k - 1).$$

By assumption $c_k - 1 > 0$ and hence the $F(p,Q,t)$ is strictly smaller than F_o, yielding a contradiction and proving the lemma.

Corollary 4.2.1 and Theorem 4.2.1 immediately yield a nonparametric variational description of $R(D)$.

Corollary 4.2.3

$$R(D) = \max_{s\le 0} \min_q (sD - \sum_j p_j \log \sum_l q_l e^{sd(j,l)}) \tag{4.2.16}$$

where the minimum is over all pmf's q.

The above variational description of $R(D)$ describes it as a maxi-min problem. We next provide an alternative characterization in terms of a maximum over s (as above), but which now involves a double maximization. This form is useful for finding lower bounds to $R(D)$.

Corollary 4.2.4

$$R(D) = \max_{s\le 0} \max_{\lambda\in\Lambda_s} (sD + \sum_j p_j \log \lambda_j) \tag{4.2.17}$$

where Λ_s is the collection of all $\lambda = \{\lambda_j\}$ such that λ_j are nonnegative and

$$c_k \equiv \sum_j p_j \lambda_j e^{sd(j,k)} \le 1, \text{ all } k.$$

Proof: The result will follow if we can prove that

$$\alpha_s \equiv \max_{\lambda\in\Lambda_s} \sum_j p_j \log \lambda_j = \min_q (\sum_j p_j \log \frac{1}{\sum_l q_l e^{sd(j,l)}}) \equiv \beta_s.$$

First suppose that q solves the right-hand minimization and hence Corollaries 4.2.3 and 4.2.2 and Lemma 4.2.1 apply. If we then define

$$\lambda_j = \frac{1}{\sum_l q_l e^{sd(j,l)}}$$

then from Lemma 4.2.1

$$c_k = \sum_j p_j \lambda_j e^{sd(j,k)} = \sum_j \frac{p_j e^{sd(j,k)}}{\sum_l q_l e^{sd(j,l)}} \leq 1$$

and hence $\lambda \in \Lambda_s$. This implies that $\alpha_s \geq \beta_s$. To derive the converse inequality, suppose that $Q \in \mathcal{Q}_{D_s}$ yields $R(D_s)$ as in Corollary 4.2.1 whence

$$R(D_s) - sD_s = \beta_s = I(Q) - s\bar{d}(Q).$$

Since $R(D_s) = I(Q)$ we have for any $s \leq 0$ and any $\lambda \in \Lambda_s$

$$\beta_s - \sum_j p_j \log \lambda_j = I(Q) - sD_s - \sum_j p_j \log \lambda_j$$

$$= I(Q) - s\bar{d}(Q) - \sum_j \sum_k p_j Q_{k|j} \log \lambda_j = \sum_j p_j \sum_k Q_{k|j} \log \frac{Q_{k|j}}{q_k \lambda_j e^{sd(j,k)}}.$$

$$= \sum_j \sum_k p_j Q_{k|j} \log \frac{p_j Q_{k|j}}{p_j q_k \lambda_j e^{sd(j,k)}} \geq 1 - \sum_{j,k} p_j q_k \lambda_j e^{sd(j,k)}$$

$$= 1 - \sum_k q_k c_k \geq 0.$$

Thus for any $s \leq 0$

$$\beta_s \geq \max_{\lambda \in \Lambda_s} \sum_j p_j \log \lambda_j = \alpha_s,$$

which completes the proof.

The previous corollary has a variation that is sometimes useful. If we replace the parameter λ_j by a parameter $f_j = \lambda_j p_j$ we get the following.

Corollary 4.2.5

$$R(D) = \max_{s \leq 0} \max_{f \in \mathcal{F}_s} (H(X) + sD + \sum_j p_j \log f_j), \tag{4.2.18}$$

where $H(X)$ is the entropy of the random variable and the maximization is over the set $\mathcal{F}_s$ of all $f = \{f_j\}$ that are nonnegative and satisfy

$$\sum_j f_j e^{sd(j,k)} \leq 1 \text{ all } k.$$

The remaining issue is how to solve the variational equations. In parametric form, this reduces to the problem of solving any of the equivalent extremum problems

$$\min_q\left(-\sum_j p_j \log \sum_l q_l e^{sd(j,l)}\right) \tag{4.2.19}$$

or

$$\max_{\lambda \in \Lambda_s} \sum_j p_j \log \lambda_j \tag{4.2.20}$$

or

$$\max_{f \in F_s} \sum_j p_j \log f_j, \tag{4.2.21}$$

since the resulting minimum or maximum together with the minimizing q or maximizing λ or f together yield the corresponding point (R_s, D_s) on the $R(D)$ curve. We now combine several previous results with one simple manipulation to provide necessary and sufficient conditions for an optimal solution. These conditions are called the *Kuhn-Tucker conditions* in optimization theory.

Theorem 4.2.3 *Necessary and sufficient conditions for a pmf q on $\hat{A}$ to solve the minimization of (4.2.16) or (4.2.19) and hence give a point on the $R(D)$ curve are that q satisfy the following equations:*

$$\sum_j p_j \frac{e^{sd(j,k)}}{\sum_l q_l e^{sd(j,l)}} = 1 \text{ if } q_k \neq 0 \tag{4.2.22}$$

$$\sum_j p_j \frac{e^{sd(j,k)}}{\sum_l q_l e^{sd(j,l)}} \leq 1 \text{ if } q_k = 0. \tag{4.2.23}$$

Proof:

We have already proved the necessity of the conditions in Corollary 4.2.4 and Lemma 4.2.1. To prove sufficiency suppose that one has a q for which the conditions are satisfied. Since $R_s - sD_s$ is the minimum over all q of

$$\beta_s(q) = \sum_j p_j \log \frac{1}{\sum_l q_l e^{sd(j,l)}}$$

we have that

$$\beta_s(q) \geq R_s(D) - sD_s. \tag{4.2.24}$$

Letting $\lambda_j = 1/\sum_l q_l e^{sd(j,l)}$ we have from Lemma 4.2.1 that $\lambda \in \Lambda_s$ and hence that

$$\beta_s(q) = \sum_j p_j \log \lambda_j = \alpha_s(\lambda) \leq \max_{\lambda \in \Lambda_s} = R_s - sD_s,$$

which with (4.2.24) implies that $\alpha_s(q) = R_s - sD_s$, which completes the proof.

Eq. (4.2.22) can be thought of as guaranteeing that the summand is a valid pmf for any fixed output k which has nonzero probability under q. Similarly, necessary and sufficient conditions on f to solve the maximization of (4.2.18) or (4.2.21) are that there exist a pmf q for which

$$f_j \sum_l q_l e^{sd(j,l)} = p_j$$

such that

$$\sum_j f_j e^{sd(j,k)} = 1 \text{ if } q_k \neq 0$$

$$\sum_j f_j e^{sd(j,k)} \leq 1 \text{ if } q_k = 0.$$

4.3 The Discrete Shannon Lower Bound

We now make a simplifying assumption that will yield a useful bound as well as complete evaluations in certain examples. It is simplest at this point to use the representation of Corollary 4.2.5, although the result can be derived from the theorem. Suppose that the input and reproduction alphabets have the same size and that distortion measure has the property that for all s

$$\sum_i e^{sd(i,k)}$$

does not depend on k and hence can be considered as a function of s only, say $\gamma(s)$. This is the case, for example, with any distortion such that the columns in the distortion matrix $\{d(j,k)\}$ are permutations of each other. It is also the case if $A = \hat{A}$ is a group and $d(j,k) = d(j-k)$ is a difference distortion measure with respect to the group difference operation. In this case it is easy to find an $f \in F_s$ and thereby obtain a lower bound to $R(D)$: simply define

$$f_j = \frac{1}{\gamma(s)} = \frac{1}{\sum_i e^{sd(i,k)}}$$

and the conditions are met and hence

$$R_s \geq sD + H(p) - \log \gamma(s). \tag{4.3.1}$$

This will hold with equality if and only if there is an output pmf q such that

$$\sum_l q_l e^{sd(j,l)} = \frac{p_j}{f_j}.$$

In matrix notation, if $E_s = 1/\gamma(s)\{e^{sd(j,k)}\}$ and p and q denote the corresponding vectors, then

$$p = E_s q.$$

Since the input and reproduction alphabets have the same size, the matrix E_s is square and the bound will hold with equality if and only if

$$q = E_s^{-1} p \tag{4.3.2}$$

is a valid pmf, that is, has all nonnegative entries. The entries automatically sum to 1 since

$$\gamma(s) \sum_k q_k = \gamma(s) \sum_j p_j.$$

This will be the case at least as $s \to -\infty$ if the distortion measure satisfies $d(j,k) = 0$ if and only if $j = k$ since then the matrix tends to the identity matrix.

The lower bound of (4.3.1) is a version of the *Shannon lower bound* and we summarize it formally for reference.

Lemma 4.3.1 *The Shannon Lower Bound*
If the distortion measure is such that

$$\gamma(s) = \sum_i e^{sd(i,k)}$$

does not depend on k, then

$$R_s \geq sD + H(p) - \log \gamma(s).$$

Furthermore, the bound will hold with equality if and only if the vector q in (4.3.2) has all nonnegative entries.

The Shannon lower bound is usually considered for difference distortion measures, that is, where the reproduction alphabet is a subset of the input alphabet and $d(j,k)$ is a function of a suitably defined difference $k - j$, say

$\rho(k-j)$. The Hamming distance and the Lee (or circular) distance are examples. In this case the condition is met and

$$\gamma(s) = \sum_a e^{s\rho(a)}.$$

The bound can often easily be optimized using calculus or standard inequalities and it sometimes yields the rate-distortion function or portions thereof. As an example, suppose that the input alphabet has K symbols and that d is given by the Hamming distance and hence

$$\gamma(s) = 1 + (K-1)e^s$$

yielding a lower bound

$$R(D) \geq sD + H(p) - \log(1 + (K-1)e^s).$$

The bound can be maximized using calculus or the divergence equality. In particular, it will be shown that the lower bound satisfies the following inequality and actually achieves it for an appropriate choice of s:

$$sD - \log(1 + (K-1)e^s) \leq -H_2(D) - D\log(K-1)$$

where

$$H_2(D) = -D\log D - (1-D)\log(1-D)$$

is the binary entropy function. To see this observe that the inequality is equivalent to

$$D\log\frac{D}{(K-1)e^s/(1+(K-1)e^s)} + (1-D)\log\frac{(1-D)}{1/(1+(K-1)e^s)} \geq 0$$

using the divergence inequality with equality if and only if

$$D = \frac{(K-1)e^s}{1+(K-1)e^s}$$

or, equivalently,

$$e^s = \frac{D}{(1-D)(K-1)}$$

or

$$s = \log\frac{D}{1-D} - \log(K-1).$$

Thus

$$R(D) \geq H(p) - H_2(D) - D\log(K-1).$$

This bound will hold with equality if the q vector of (4.3.2) has all of its entries nonnegative. These entries can be found to be

$$q_k = \frac{p_k[1 + (K-1)e^s] - e^s}{1 - e^s}$$

which will be nonnegative if

$$p_k \geq \frac{1}{e^{-s} + (K-1)}$$

for all k. This will hold provided s is sufficiently negative. In summary,

$$R(D) \geq H(p) - H_2(D) - D\log(K-1) \quad (4.3.3)$$

with equality for

$$0 \leq D \leq (K-1)p_{min} \quad (4.3.4)$$

where

$$p_{min} = \min_j p_j.$$

If the input is equiprobable, then the above formula gives the rate-distortion function for all D in $[0, 1 - 1/K]$. This, however, is the entire range of D since $R(1 - 1/K) = 0$. To see this observe that in the general non-equiprobable case the best zero rate code consists of only one symbol, the most probable source symbol, say the kth symbol. Using this codeword yields an average Hamming distortion of $\sum_{j \neq k} p_j = 1 - p_k$. In the equiprobable case, this is $1 - 1/K$.

4.4 The Blahut Algorithm

Theorem 4.2.2 yields a simple but powerful iterative algorithm for finding rate-distortion or distortion-rate curves. For a fixed value of the parameter s, in order to find the (D_s, R_s) point on the $R(D)$ curve we need to accomplish the minimization

$$\min_q \min_Q F(p, Q, q)$$

where $F(p, Q, q)$ is given by (4.2.4). Theorem 4.2.2 part (b) provides the optimum q for a given Q and part (c) provides the optimum Q for a given q. This suggests the following algorithm:

1. Start with an initial guess, say q^1 for q, the output distribution. Set $m = 1$.

2. Given q^m, find the optimal Q, say Q^m, for q_m. This is given by

$$Q^m_{k|j} = \frac{q^m_k e^{sd(j,k)}}{\sum_l q^m_l e^{sd(j,l)}}.$$

3. Given Q^m, find the optimal q, say q^{m+1}, for Q^m. This is given by

$$q^{m+1}_k = \sum_j p_j Q^m_{k|j}.$$

4. Evaluate $R_m = I(q^m, Q^m)$ and $D_m = D(q^m, Q^m)$ and test for approximate convergence. Either quit or set $m+1 \to m$ and go to step 1.

We will not try to prove the convergence of the algorithm, but we will briefly describe why it works. Theorem 4.2.2 (b) and (c) prove that at each step of the iteration $F(p, Q^m, q^m)$ is decreased and hence the monotone sequence of nondecreasing functions must have a limit. Results from elementary real analysis imply that some subsequence of q^m (and hence also Q^m) converges to a limiting pmf which must satisfy simultaneously conditions (b) and (c) (a stationary point of the algorithm) and also must satisfy the necessary and sufficient conditions of the lemma for achieving the minimum. This algorithm where one alternatively performs easy minimizations of an information measure with respect to two different distributions is a special case of a general algorithm of Csiszár and Tusnady [24] and the convergence properties follow from their results. It is also a special case of an algorithm called the EM algorithm.

4.5 Continuous Alphabets

When the alphabet is continuous, the calculus of variations can be used to extend the above results. Rigorous proofs of these results require considerably more care than the corresponding discrete proofs. A detailed discussion can be found in Csiszár [23].

The following theorem summarizes several of the properties, which can be viewed as the natural extension from sums to integrals.

Theorem 4.5.1 *Let X be a random variable described by a pdf p. Then*

$$R(D) = \sup_{s \leq 0} \min_q (sD - \int dx p(x) \log \int dy q(y) e^{sd(x,y)}.)$$

Necessary and sufficient conditions for an output pdf q to achieve the above minima are that it satisfy the relations

$$\int dx \frac{e^{sd(x,y)}}{\int dz q(z) e^{sd(x,z)}} = 1 \text{ if } q(y) \neq 0$$

$$\int dx \frac{e^{sd(x,y)}}{\int dz q(z) e^{sd(x,z)}} \leq 1 \text{ if } q(y) = 0.$$

We also have that

$$R(D) = \sup_{s \leq 0} \sup_{f \in \mathcal{F}_s} (h(X) + sD + \int dx p(x) \log f(x)), \tag{4.5.1}$$

where $h(X)$ is the differential entropy of the random variable and the maximization is over all $f = f(x)$ that are nonnegative and satisfy

$$\int dx f(x) e^{sd(x,y)} \leq 1 \text{ for all } y.$$

Necessary and sufficient conditions for the f to achieve the supremum are that there exists a pdf q such that

$$f(x) \int dy q(y) e^{sd(x,y)} = p(x)$$

such that

$$\int dx f(x) e^{sd(x,y)} = 1 \text{ if } q(y) \neq 0$$

$$\int dx f(x) e^{sd(x,y)} \leq 1 \text{ if } q(y) = 0.$$

4.6 The Continuous Shannon Lower Bound

The Shannon lower bound is most useful for continuous alphabets, in which case a difference distortion measure is almost always assumed. The following development is similar to Berger [5], except that (as usual) we prove optimality via the divergence inequality (Lemma 2.6.1) instead of using calculus to investigate first and second derivatives.

Recall that difference distortion measures have the general form

$$d(x, y) = L(x - y).$$

We assume that the distortion measure is such that

$$\gamma(s) = \int dy e^{sL(y)} < \infty.$$

In this case we can use the theorem of the previous section or mimic the discrete development of the Shannon lower bound to obtain the bound. The arguments of the discrete case can be replaced by the corresponding continuous arguments, replacing the pmf's by pdf's and the entropy by the differential entropy to obtain the parametric bound

$$R(D) \geq h(p) + sD - \log \gamma(s). \tag{4.6.1}$$

Define the pdf

$$g_s(y) = \frac{e^{sL(y)}}{\int dz e^{sL(z)}}.$$

The best lower bound of the form (4.6.1) is obtained by choosing s to maximize the lower bound. We claim that the globally optimum s is given by a value, say s^*, which solves the equation

$$\int L(y) g_{s^*}(y)\, dy = D. \tag{4.6.2}$$

To prove this claim we need to show that for all s

$$s^* D - \log \gamma(s^*) \geq sD - \log \gamma(s).$$

We have that

$$\begin{aligned} & s^* D - \log \gamma(s^*) - (sD - \log \gamma(s)) \\ &= \int dy g_{s^*}(y) \left(s^* L(y) - \log \gamma(s^*) - sL(y) + \log \gamma(s) \right) \\ &= \int dy g_{s^*}(y) \log \frac{g_{s^*}(y)}{g_s(y)} \geq 0, \end{aligned}$$

using the continuous version of the divergence inequality.

As an aid to intuition, note that differentiating (4.6.1) with respect to s yields (4.6.2), showing that such an s is at least a stationary point.

When s is selected in this manner, the bound can be written as

$$R_{\text{SLB}}(D) = h(p) - h(g_s). \tag{4.6.3}$$

We summarize the bound in the following lemma, where we omit the parameter s by replacing s by $-b(D)$, showing that the optimizing s of (4.6.2) depends on D, and setting $a(D) = 1/\gamma(s)$.

Lemma 4.6.1 *Given a continuous random variable described by a pdf p and a difference distortion measure $d(x,y) = L(x-y)$, then*

$$R(D) \geq R_{\mathrm{SLB}}(D)$$

where

$$R_{\mathrm{SLB}}(D) = h(p) + \log a(D) - Db(D) \tag{4.6.4}$$

where $a(D)$ and $b(D)$ are solutions to the equations

$$a(D) \int e^{-b(D)L(x)} dx = 1 \tag{4.6.5}$$

$$a(D) \int L(x) e^{-b(D)L(x)} dx = D. \tag{4.6.6}$$

Eq. (4.6.3) can be used to derive an alternative form. We have that

$$\int g_s(y) L(y)\, dy = D.$$

The following lemma shows that the pdf g_s yields the largest differential entropy over all pdf's which have the same or smaller distortion.

Lemma 4.6.2

$$h(g_s) = \max_{g \in \mathcal{G}_D} h(g),$$

where $\mathcal{G}_{\mathcal{D}}$ is the collection of all pdf's g for which

$$\int dy g(y) L(y) \leq D.$$

Proof:
As usual, the proof is based on the divergence inequality of Lemma 2.6.1. First observe that $h(g_s) = sD - \log \gamma(s)$ and then break up $h(q)$ as

$$\int g(y) \log \frac{1}{g(y)}\, dy = -\int g(y) \log \frac{g(y)}{g_s(y)}\, dy + \int g(y) \log \frac{1}{g_s(y)}\, dy.$$

Applying the divergence inequality then yields

$$h(g) \leq 0 + \int g(y) \log \frac{e^{sL(y)}}{\gamma(s)}\, dy = s \int L(y) g(y)\, dy - \log \gamma(s)$$

$$= sD - \gamma(s) = h(g_s),$$

proving the lemma.

Thus the Shannon lower bound can also be written

$$R_{\mathrm{SLB}}(D) = h(p) - \max_{g \in \mathcal{G}_D} h(g), \tag{4.6.7}$$

This is the form originally developed by Shannon.

As an example, suppose that the distortion measure is the squared error distortion $L(y) = y^2$. Then

$$\int e^{-b(D)y^2}\, dy = \left(\frac{\pi}{b(D)}\right)^{1/2}$$

so that

$$a(D) = \left(\frac{b(D)}{\pi}\right)^{1/2}$$

and hence the second requirement becomes

$$\left(\frac{b(D)}{\pi}\right)^{1/2} \int y^2 e^{-b(D)y^2}\, dy = \frac{1}{2b(D)} = D.$$

Thus

$$R_{\mathrm{SLB}}(D) = h(p) - \frac{1}{2}\log 2\pi D - \frac{1}{2}.$$

If we use natural logarithms this becomes

$$R_{\mathrm{SLB}} = h(p) - \frac{1}{2}\log 2\pi e D.$$

If in addition the source is Gaussian with 0 mean and variance σ^2, then

$$h(p) = \frac{1}{2}\log 2\pi e \sigma^2$$

and hence

$$R_{\mathrm{SLB}}(D) = \frac{1}{2}\log\frac{\sigma^2}{D}, \tag{4.6.8}$$

which bound is nonnegative for $D \in [0, \sigma^2]$. It can be shown using Theorem 4.5.1 that this is in fact the rate-distortion function for the given region. Note that we can easily invert $R(D)$ in this case to find the distortion-rate function

$$D(R) = \sigma^2 e^{-2R}.$$

4.7 Vectors and Processes

The previous techniques can be extended in a straightforward manner to finite-dimensional vectors. All of the properties and variational descriptions have their multidimensional analogs. The Blahut algorithm also generalizes without difficulty. We here consider several such generalizations that are useful for evaluating and bounding multidimensional rate-distortion functions.

We begin with given a k dimensional random vector X^k and a distortion measure $d(x^k, y^k)$ giving the distortion resulting between vectors. Then one can define a rate-distortion function for the vector in exactly the same way as before, that is,

$$R_{X^k}(D) = \inf_{p_{Y^k|X^k}: Ed(X^k,Y^k) \leq D} I(X^k; Y^k),$$

where we have subscripted the rate-distortion function with the name of the random vector. In this form all of the previous variational expressions and bounds have immediate generalizations. This form is somewhat clumsy, however, when our real goal is to normalize both rate and distortion and to consider the limiting rate and distortion per symbol. For this reason we often instead focus on the kth order rate-distortion function $R_k(D)$ of a process formed by normalizing the rae-distortion function of the vector. The relation between these two quantities depends on how we define the distortion between vectors. Suppose that $d_k(x^k, y^k)$ is an unnormalized vector distortion as previously considered, e.g.,

$$d_k(x^k, y^k) = \sum_{i=0}^{k-1} d_1(x_i, y_i)$$

for an additive fidelity criterion. Then if we define

$$d(x^k, y^k) = d_k(x^k, y^k),$$

that is, an unnormalized distortion, then

$$R_k(D) = \frac{1}{k} R_{X^k}(kD).$$

If, on the other hand, we normalize the distortion and define

$$d(x^k, y^k) = \frac{1}{k} d_k(x^k, y^k),$$

then

$$R_k(D) = \frac{1}{k} R_{X^k}(D).$$

Regardless of which way we define the vector distortion, both formulas imply that

$$R_k(D) = \frac{1}{k} \inf_{p_{Y^K|X^K}} I(X^k; Y^k),$$

where the infimum is over all conditional pdf's satisfying the condition

$$\int dx^k dy^k p_{Y^k|X^k}(y^k|x^k) p_{X^k}(x^k) d_k(x^k, y^k) \le kD.$$

For a fixed integer k, the k dimensional version of (4.5.1) is given by the following theorem:

Theorem 4.7.1 *Let $R_k(D)$ be the kth order rate-distortion function. Let p_k denote the pdf p_{X^k}, $h(p_k)$ denote the differential entropy*

$$h(p_k) = \int dx^k p_k(x^k) \log \frac{1}{p_k(x^k)},$$

and

$$d(x^k, y^k) = \frac{1}{k} d_k(x^k, y^k)$$

denote the vector distortion per symbol. Then

$$kR_k(D) = h(p_k) + \sup_{s \le 0} \sup_{f \in \mathcal{F}_s} \left(sD + \int p_k(x^k) \log f(x^k) dx^k \right) \tag{4.7.1}$$

where $\mathcal{F}_s$ is the set of all $f = f(x^k)$ that are nonnegative and satisfy

$$\int dx^k f(x^k) e^{s d(x^k, y^k)} \le 1, \text{ all } y^k. \tag{4.7.2}$$

4.7.1 The Wyner-Ziv Lower Bound

When the source is stationary and the distortion measure is additive, that is, when

$$d(x^k, y^k) = \frac{1}{k} \sum_{i=0}^{k-1} d_1(x_i, y_i),$$

a simple bound to the kth order rate-distortion function can be found in terms of the first order bound as follows: Suppose that s and $f_1(x)$ yield the first order rate-distortion function (solves the $k = 1$ case above) and hence

$$\int dx_1 f_1(x_1) e^{s d_1(x_1, y_1)} \le 1, \text{ all } y_1,$$

$$R_1(D) = h(p_1) + sD + \int p_1(x_1) \log f_1(x_1)\, dx_1.$$

Define the function

$$f(x^k) = \prod_{i=0}^{k-1} f_1(x_i).$$

Then choosing $s' = sk$ we have that

$$\int dx^k f(x^k) e^{s' d(x^k, y^k)} = \prod_{i=0}^{k-1} \int dx_i f_1(x_i) e^{s d_1(x_i, y_i)} \leq 1,$$

and hence $f \in \mathcal{F}_{sk}$. Thus

$$kR_k(D) \geq h(p_k) + s'D + \int p_k(x^k) \log \prod_{i=0}^{n-1} f(x_i)\, dx^k$$

$$= h(p_k) + s'D + \sum_{i=0}^{k-1} \int dx_i p_1(x_i) \log f(x_i)$$

$$= h(p_k) + s'D + \sum_{i=0}^{k-1} (R_1(D) - h(p_1) - sD) = h(p_k) + k(R_1(D) - h(p_1)).$$

Thus

$$R_k(D) \geq R_1(D) - (h(p_1) - \frac{1}{k} h(p_k)), \tag{4.7.3}$$

which points out that the kth order rate-distortion function is bound below by the first order rate-distortion function minus a term that depends only on the source and *not* on the distortion measure. Note that $R_1(D)$ can also be interpreted as the rate-distortion function of a memoryless source having the same marginals as the original source. If the limits exist, we can take the limit as $k \to \infty$ to find a bound for the rate-distortion function of the process:

$$R(D) \geq R_1(D) - (h(p_1) - \lim_{k \to \infty} \frac{1}{k} h(p_k))$$

$$= R_1(D) - (h(X_0) - \bar{h}(X)), \tag{4.7.4}$$

where $\bar{h}$ is the differential entropy rate and where we have used the notation emphasizing the names of the random process and variables rather than the distribution. These bounds are originally due to Wyner and Ziv [83], but the proof is closer to Berger's [5]. A similar bound holds for discrete alphabet processes with entropy replacing differential entropy.

4.7.2 The Autoregressive Lower Bound

Another lower bound can be constructed in a similar fashion in the special case where the distortion measure is again an additive difference distortion measure and the source is an autoregressive source: Suppose that Z_n is an iid source with rate-distortion function $R_Z(D)$ and suppose that X_n is an autoregressive source defined by

$$X_n = -\sum_{k=1}^{n} X_{n-k}a_k + Z_n;\ n = 0, 1, 2, \ldots.$$

We assume that the a_k are square summable so that the above process will be asymptotically stationary. Suppose moreover that f_Z and s yield the supremum of the theorem and yield $R_Z(D)$. Define for k

$$f(x^k) = \prod_{i=0}^{k-1} f_Z(x_i + \sum_{j=1}^{i} a_j x_{i-j}). \tag{4.7.5}$$

Observe that like the Wyner-Ziv bound the function f is constructed as a product, but here the arguments of the terms in the product are different. It can be shown by changing variables of integration that f meets the conditions required by the maximization describing the kth order rate distortion function for $\{X_n\}$. This then yields from the theorem the bound

$$kR_k(D) \geq$$

$$h(p_{X^k}) + ksD + \int dx^k \prod_{i=0}^{k-1} p_Z(x_i + \sum_{j=0}^{i-1} a_j x_{i-j}) \log \prod_{i=0}^{k-1} f_Z(x_i + \sum_{j=0}^{i-1} a_j x_{i-j}).$$

By repeated changes of variables it can be seen that

$$h(p_{X^k}) = kh(p_Z) \tag{4.7.6}$$

and that the remaining terms reduce to

$$k(sD + \int dz p_Z(z) \log f_Z(z))$$

and hence

$$R_k(D) \geq R_Z(D).$$

Since this is true for all k, taking the limit yields

$$R(D) \geq R_Z(D), \tag{4.7.7}$$

which is called the *autoregressive lower bound*. Like the Wyner-Ziv bound, the autoregressive lower bound bounds $R(D)$ for a process with memory by that of a related memoryless process. Here, however, the related process is the memoryless innovations process which produces the source by an autoregressive filter and not the memoryless source having the same marginals. A version of the bound also exists for discrete alphabet processes when a difference distortion measure is used [35].

4.7.3 The Vector Shannon Lower Bound

We next consider a bound that does not require additive distortion measures, but does assume difference distortion measures. When the bound for vectors is used to obtain a bound for processes, however, we will once again assume additive distortion measures. The bound is the vector generalization of the Shannon lower bound and the proof is virtually unchanged from the scalar case. Hence the general result is stated in the following lemma without proof. Note that in this bound we consider the ordinary rate distortion function without normalizing either rate or distortion.

Lemma 4.7.1 *Suppose that X is a k-dimensional random vector with a pdf p_k and that we are given an (unnormalized) difference distortion measure $d(x,y) = L(x-y)$, where the difference is an ordinary Euclidean distance. Define $R_X(D)$ by*

$$R_X(D) = \inf I(X;Y)$$

where the infimum is over all conditional probability measures for which

$$Ed(X,Y) \leq D.$$

Then

$$R_X(D) \geq R_{SLB}(D)$$

where

$$R_{SLB}(D) = h(p_k) + \log a(D) - Db(D)$$

where $a(D)$ and $b(D)$ are solutions to the equations

$$a(D) \int e^{-b(D)L(x)} dx = 1$$

$$a(D) \int L(x) e^{-b(D)L(x)} dx = D,$$

and where $h(p_k)$ is the differential entropy of the pdf p_k.

Note that in order to relate the rate-distortion function $R(D)$ of the vector as defined above to the kth order rate-distortion function previously considered, we need to normalize both rate and distortion, that is,

$$R_k(D) = \frac{1}{k} R(kD).$$

The Shannon lower bound yields explicit lower bounds in the special case of distortion measures that depend on norms in Euclidean space. Hence in the next section we develop some of the basic properties of such distortion measures and the corresponding bounds.

4.8 Norm Distortion

In this section we derive the Shannon lower bound for norm-based distortion measures. This provides both specific examples of the evaluation of Shannon lower bounds and the most general case where such bounds are known. The general bounds were first developed for Euclidean norms by Lin'kov [59] and were subsequently corrected and generalized to arbitrary norms by Yamada et al. [84].

We assume that d is a distortion measure on k-dimensional Euclidean space with the form

$$d(x, y) = \rho(||x - y||)$$

for some norm $||\cdot||$ and some function ρ which is nondecreasing in its argument. The vector difference is ordinary Euclidean difference. The special case of principal interest in this section is $\rho(\alpha) = \alpha^r$ for $r \geq 1$ and we henceforth assume that

$$d(x, y) = ||x - y||^r.$$

An example is the pth power distortion measure

$$d(x, y) = \sum_{i=0}^{k-1} |x_i - y_i|^p = ||x - y||_p^p$$

and hence also the squared error distortion measure.

The bounds for norm-based distortion measures depend on the volume of a unit sphere with respect to the norm. Given a set $G \subset \mathcal{R}^k$, define its volume $V(G)$ by the integral

$$V(G) = \int_G dx.$$

The unit sphere with respect to a seminorm is the set $\{x : ||x|| \leq 1\}$. Let V_k denote the volume of the unit sphere in k dimensional space with respect to a given seminorm

$$V_k = \int_{||x|| \leq 1} dx. \tag{4.8.1}$$

This volume is computable for most interesting seminorms. For example, for the l^p norms

$$V_k = \frac{2^k (\Gamma(1/p))^k}{k\Gamma(k/p)p^{k-1}}, \tag{4.8.2}$$

where Γ is the gamma function

$$\Gamma(n) = \int_0^\infty e^{-u} u^{n-1} du.$$

(See Gradshteyn and Ryzhik [34], p. 620.) For example, if $p = 2$ we can use the fact that $\Gamma(1/2) = \sqrt{\pi}$ to find

$$V_k = \frac{2\pi^{k/2}}{k\Gamma(k/2)}. \tag{4.8.3}$$

For the l^∞ norm it is easily seen that $V_k = 1$. For the weighted quadratic norm $||x|| = x^t \mathbf{B} x$ with a positive definite matrix $\mathbf{B}$, the above result for l^2 combined with some matrix theory and a change of variables yields

$$V_k = (\det \mathbf{B})^{-1/2} \frac{2\Gamma(1/2)^k}{k\Gamma(k/2)}, \tag{4.8.4}$$

where $\Gamma(1/2) = \sqrt{\pi}$.

Applying the vector Shannon lower bound of Lemma 4.7.1 to seminorm-based distortion measures using a change of variables and integral tables the integrals can be evaluated to find that

$$a(D) = \frac{r}{kV_k} \frac{b(D)^{k/r}}{\Gamma(k/r)} \tag{4.8.5}$$

$$b(D) = \frac{k}{rD} \tag{4.8.6}$$

and hence

$$R_{\mathrm{SLB}}^{(k)}(D) = h(p_k) - \frac{k}{r} + \log\left(\frac{r}{kV_k}\left(\frac{k}{rD}\right)^{k/r} \frac{1}{\Gamma(\frac{k}{r})}\right), \tag{4.8.7}$$

where V_k is the volume of the unit sphere with respect to the given norm. Observe that the only dependence on the particular norm used is through V_k.

Normalizing the rate and distortion by $1/k$ we have (using base e logarithms)

$$\frac{1}{k}R_{SLB}^{(k)}(D) = \frac{1}{k}h(p_k) - \frac{1}{r}\log\frac{reD}{k} - \frac{1}{k}\log\frac{V_k k}{r}\Gamma(\frac{k}{r})$$

and hence we have the following bound for the kth order rate-distortion function:

$$R_k(D) \geq \frac{1}{k}h(p_k) - \frac{1}{r}\log reD - \frac{1}{k}\log\frac{V_k k}{r}\Gamma(\frac{k}{r}). \tag{4.8.8}$$

Note that D has replaced D/k since the average distortion is so normalized in the kth order rate-distortion function. We can also write this as a distortion rate function: Defining the normalized differential entropy $h_k = h(p_k)/k$

$$\begin{aligned} D_k(R) &\geq \frac{e^{r(h_k-R)}(\frac{V_k k}{r}\Gamma(k/r))^{-r/k}}{re} \\ &= \frac{(V_k\Gamma(1+\frac{k}{r}))^{-\frac{r}{k}}}{re}e^{-r(R-h_k)} \end{aligned} \tag{4.8.9}$$

Thus if the l^p norm is used,

$$R_k(D) \geq h_k - \frac{1}{r}\log reD - \frac{1}{k}\log\frac{2^k(\Gamma(\frac{1}{p}))^k}{\Gamma(\frac{k}{p})p^{k-1}}\Gamma(\frac{k}{r}). \tag{4.8.10}$$

If $p = r$ (and hence the distortion measure is the additive pth power distortion measure)

$$R_k(D) \geq h_k - \frac{1}{r}\log reD - \frac{1}{k}\log\frac{2^k(\Gamma(\frac{1}{p}))^k}{p^{k-1}}.$$

These vector bounds can be used to derive lower bounds to the process rate-distortion function: For the general case of the rth power of an arbitrary norm we have that

$$R(D) = \lim_{k\to\infty} R_k(D) \geq \bar{h}(X) - \frac{1}{r}\log reD - \lim_{k\to\infty}\frac{1}{k}\log V_k k\Gamma(\frac{k}{r})$$

where $\bar{h}(X)$ is the differential entropy rate of the process

$$\bar{h}(X) = \lim_{k\to\infty} h_k = \lim_{k\to\infty}\frac{1}{k}h(p_k).$$

In the case of l^p norms this becomes

$$\bar{h}(X) - \frac{1}{r}\log reD - \log\frac{2\Gamma(\frac{1}{p})}{p} - \lim_{k\to\infty}\frac{1}{k}\log p\frac{\Gamma(\frac{k}{r})}{\Gamma(\frac{k}{p})}.$$

If $r = p$ and hence the distortion measure is just the rth power (and hence additive), we have immediately that

$$R(D) \geq \bar{h}(X) - \frac{1}{r}\log reD - \log\frac{2\Gamma(\frac{1}{r})}{r}. \tag{4.8.11}$$

If $p \neq r$, it can be shown using Stirling's approximation that the limit diverges. Observe that if $r = 2$, then $\Gamma(1/2) = \sqrt{\pi}$ and the bound reduces to

$$R(D) \geq \bar{h}(X) - \frac{1}{2}\log 2\pi eD. \tag{4.8.12}$$

As an example, suppose that $\{X_n\}$ is a Gaussian random process with zero mean, autocorrelation function $R_X(k)$, and power spectral density $S_X(f)$. Let R_k denote the $k \times k$ autocorrelation matrix $R_X = E(XX^t)$, where $X^t = (X_0, \ldots, X_{k-1})$. Then

$$h(p_k) = -\int_{\mathcal{R}^k} dx \frac{e^{-\frac{1}{2}x^t R_k^{-1} x}}{\sqrt{(2\pi)^k \det R_k}} \log\frac{e^{-\frac{1}{2}x^t R_k^{-1} x}}{\sqrt{(2\pi)^k \det R_k}}$$

$$= \frac{k}{2}\log(2\pi) + \frac{1}{2}\log\det R_k + \frac{1}{2}\int dx p_k(x) x^t R_k^{-1} x.$$

The rightmost term can be expressed as

$$E(X^t R_k^{-1} X) = E(\operatorname{Tr} R_k^{-1} XX^t) = \operatorname{Tr}(R_k^{-1} E(XX^t))$$

$$= \operatorname{Tr}(R_k^{-1} R_k) = \operatorname{Tr} I = k,$$

where "Tr" denotes the trace of a matrix (the sum of its diagonal elements), the representation of the quadratic form as the given trace can be verified by direct substitution, and I denotes the identity matrix. Thus

$$\frac{1}{k}h(p_k) = \frac{1}{2}\log(2\pi e) + \frac{1}{2k}\log\det R_k.$$

The Gaussian is one of the few examples where the differential entropy of vectors can be exactly evaluated. The differential entropy rate can be found by taking the limit using a fundamental result of Toeplitz matrices (see, e.g., [44] or [36]):

$$\lim_{k\to\infty}\frac{1}{k}\log\det R_k = \int_{-\frac{1}{2}}^{\frac{1}{2}}\log S_X(f)\,df = \log\sigma_\epsilon^2$$

where σ_ϵ^2 is called the *one-step prediction error* of the process. (See Chapter 6 in [30] for a derivation of the one-step prediction error using the theory of linear prediction.) Thus in the case of a Gaussian process, we have an explicit representation of the differential entropy rate

$$\bar{h} = \frac{1}{2}\log(2\pi e) + \frac{1}{2}\int_{-\frac{1}{2}}^{\frac{1}{2}} \log S_X(f)\,df = \frac{1}{2}\log(2\pi e\sigma_\epsilon^2). \tag{4.8.13}$$

This in turn provides an explicit representation of the asymptotic Shannon lower bound for a Gaussian process with a squared-error distortion:

$$R(D) \geq R_{\text{SLB}}(D) = \frac{1}{2}\log(\frac{\sigma_\epsilon^2}{D}). \tag{4.8.14}$$

Note the similarity with the Shannon lower bound for a single Gaussian random variable of (4.6.8). The variance of the single random variable has been replaced by the one-step prediction error.

Although the good news is that we have an explicit formula for the Shannon lower bound of a Gaussian process with a squared error distortion, the drawback is that for the special case of a Gaussian process the rate-distortion function can be explicitly evaluated, although the details are somewhat tedious. The interested reader can pursue the subject in Berger [5] and Gallager [28].

Exercises

1. Find the Shannon lower bound to the rate-distortion function of a zero mean iid Gaussian random process with a magnitude error distortion measure $d(x, y) = |x - y|$. Does the Shannon lower bound to $R_1(D)$ actually yield $R_1(D)$ for any value of D?

2. Show that for a finite alphabet source and a distortion measure with the property that $d(k, k) = 0$, all k, it is true that

$$\lim_{D\to 0}(R_1(D) - R_{\text{SLB}}(D)) = 0,$$

where $R_1(D)$ is the first order rate-distortion function and $R_{\text{SLB}}(D)$ is its Shannon lower bound.

3. Strengthen the previous problem by showing that in this case there is a $D_{\text{crit}} > 0$ with the property that

$$R_1(D) = R_{\text{SLB}}(D);\ 0 \leq D \leq D_{\text{crit}}.$$

4. Given two discrete random variables X, Y define the *conditional rate-distortion function* $R_{X|Y}(D)$ for X given Y by
$$R_{X|Y}(D) = \inf_{p_{\hat{X}|X,Y}} I(X;\hat{X}|Y),$$
where the infimum is over all conditional pmfs $p_{\hat{X}|X,Y}$ for which
$$Ed(X;\hat{X}) \leq D.$$
Prove that
$$R_X(D) \geq R_{X|Y}(D) \geq R_X(D) - I(X;Y).$$

5. Develop the Wyner-Ziv bound for discrete alphabet stationary sources. Evaluate the bound for the case of a binary Markov source; that is, the source X_n defined by
$$X_n = X_{n-1} + Z_n,$$
where $\{Z_n\}$ is a binary iid source with $\Pr(Z_n = 1) = 1 - \Pr(Z_n = 0) = p$ and where the addition is modulo 2. Develop an autoregressive lower bound for the same source and compare the two bounds.

6. Prove that f in (4.7.5) meets the conditions of (4.7.2).

7. Prove (4.7.6).

8. Evaluate the Shannon lower bound of (4.8.5 –4.8.7) for the case of a Gauss Markov source $\{X_n\}$ with 0 mean and autocorrelation function $R_X(k) = EX_0X_k = \sigma_X^2\rho^{|k|}$, all k, and a squared error distortion measure. Repeat for a magnitude error distortion measure.

9. Verify (4.8.4).

10. Verify (4.8.5 -4.8.7).

11. Show that for an iid process,
$$R_N(D) = R_1(D) = R(D).$$

12. Suppose that $\{X_n\}$ is an iid process with a marginal pdf that is uniform on the interval $[-a, a]$. Evaluate or bound the rate-distortion function for this process, $R_X(D)$, for a squared error and a magnitude error distortion measure. Repeat for a Gaussian process $\{Y_n\}$ that is also iid and has the same mean and variance as $\{X_n\}$ and has rate-distortion function $R_Y(D)$. Compare $R_X(D)$ and $R_Y(D)$ and comment on the difference.

13. Show that for an arbitrary discrete alphabet stationary source

$$R_1(D) \geq R(D).$$

14. Suppose that $\{X_n\}$ is an autoregressive Gaussian process. Show that the autoregressive lower bound also yields the bound (4.8.14).

15. Write a program for the Blahut algorithm and try it out on a Poisson source with a Hamming distortion measure. You may wish to read the original Blahut development [6] in order to get a more efficient stopping rule for the algorithm.

Chapter 5

High Rate Quantization

5.1 Introduction

In this chapter we find bounds and approximations to the performance of memoryless vector quantizers or block source codes under the assumption that the rate of the code is high (and hence the distortion is low). These approximations are also referred to as asymptotic quantization approximations (asymptotic in the sense of large rate) and fine quantization. This approach contrasts with the usual Shannon theory in that we do not explicitly define an information theoretic optimization and then relate it to the performance of deterministic codes via a coding theorem. Instead we assume a certain asymptotic structure to a deterministic code and use basic calculus approximations to find the resulting distortion, rate, and entropy. The results so obtained do relate to the Shannon rate distortion functions and we close with some such comparisons. The theory will be presented in a less rigorous style than the Shannon theory because being precise would require an unpleasant amount of detailed integration theory.

The theory of high rate quantization was first developed for scalar quantizers by Bennett [4] and subsequently extended to vector quantizers by Schutzenberger [72] and Zador [85], [86], [87]. These results were later generalized by Gish and Pierce [32], Gersho [29], Yamada et al. [84], and Bucklew and Wise [11]. The latter paper provides rigorous derivations of the basic results for rth power distortion measures.

The information source $\{X_n\}$ is assumed to be continuous with probability density functions that are well-behaved in the sense that we can approximate multidimensional integrals by sums and vice-versa. In addition, we confine interest to a special class of distortion measures, the seminorm-based distortion measures used in the examples of the Shannon lower bound. This is unfortunately not just for convenience; the techniques used do not work for very general distortion measures.

We now consider the coding of source vectors of fixed dimension, say k. As k is fixed and as vectors will have the same dimension, we simplify notation by writing random vectors as upper case italics and sample vectors as lower case italics, e.g., we use x instead of x^k or boldface $\mathbf{x}$.

Let X be a k-dimensional real valued random vector described by a pdf f (we omit the subscript X on the pdf as X is the only random vector to be considered). We will later place constraints on f, but basically we require that it be sufficiently "nice" for the fundamental theorem of calculus to hold, that is, that integrals of f over small regions can be approximated by the volume of the region times a value of f inside the region. This is the basic assumption that we shall use repeatedly.

For an integer N let q be a mapping from k-dimensional Euclidean space $\mathcal{R}^k$ into a reproduction alphabet $\hat{A}_N$ having N elements $y_1, \ldots, y_N$. If $q(X) = y_i$, then the index i is the channel codeword and y_i is the decoded reproduction. Define the encoder partition corresponding to the quantizer as the collection of sets S_i defined by

$$S_i = \{x : \ q(x) = y_i\},$$

that is, the collection of all input vectors mapping into the ith reproduction letter. The function q can be described as

$$q(x) = \sum_{i=1}^{N} 1_{S_i}(x) y_i,$$

where $1_{S_i}(x)$ is the indicator function of the set S_i; that is,

$$1_{S_i}(x) = \begin{cases} 1 & \text{if } x \in S_i \\ 0 & \text{otherwise.} \end{cases}$$

As in Section 4.8, we shall consider only seminorm-based distortion measures, that is, distortion measures of the form

$$d(x, y) = \rho(||x - y||),$$

where ρ is a nondecreasing function of its argument. The example of principal interest is $\rho(\alpha) = \alpha^r$ for some $r \geq 1$, the example used for the vector Shannon lower bounds.

Given a seminorm-based distortion measure, the optimum (minimum distortion) quantization regions are given by

$$S_i = \{x : \rho(||x-y_i||) \leq \rho(||x-y_j||); \text{ all } j\} = \{x : \ ||x-y_i|| \leq ||x-y_j||; \text{ all } j\}$$

since $\rho(\alpha) \leq \rho(\beta)$ if and only if $\alpha \leq \beta$ since ρ is nondecreasing in its argument. These sets are called the *Voronoi cells* or *Voronoi regions* of the quantizer. They are also called the ***nearest neighbor regions*** and the ***Dirichlet regions***.

In the special case where the Euclidean norm is used, an equivalent definition is

$$S_i = \{x : x^t y_i - \frac{1}{2}||y_i||^2 \geq x^t y_j - \frac{1}{2}||y_j||^2\}$$

and membership in the region can be determined by computation of inner products minus a bias term. When the S_i is bounded, then the region is bounded by segments of $k-1$ dimensional hyperplanes and forms a convex polytope in k-dimensional space.

5.2 Asymptotic Distortion

Given a quantizer q with output vectors $\{y_i;\ i = 1, \ldots, N\}$ and input partition $\{S_i;\ i = 1, \ldots, N\}$ and a difference distortion measure as detailed in the previous section, the average distortion of the quantizer can be expressed as

$$\begin{aligned} \bar{d} &= Ed(X, q(X)) = E\rho(||X - q(X)||) \\ &= \sum_{i=1}^{N} \int_{S_i} f(x)\rho(||x - y_i||)\, dx. \end{aligned}$$

We now assume that N is so large and the pdf f sufficiently smooth so that for all bounded cells S_i the pdf $f(x)$ is nearly constant within S_i, say given by a value f_i, and we can make the approximation

$$\int_{S_i} f(x)\rho(||x - y_i||)\, dx \approx f_i \int_{S_i} \rho(||x - y_i||)\, dx.$$

Furthermore, since $f(x)$ is approximately f_i within the cell, we can integrate to obtain $P_i = \Pr(X \in S_i)$ as

$$P_i = \int_{S_i} f(x)\, dx \approx f_i V(S_i)$$

and hence the distortion approximation can also be expressed as

$$\int_{S_i} f(x)\rho(||x - y_i||)\, dx \approx \frac{P_i}{V(S_i)} \int_{S_i} \rho(||x - y_i||)\, dx. \tag{5.2.1}$$

In order to use this approximation throughout, we assume that the total probability of the input vector falling within unbounded quantization cells is negligible. Extending traditional scalar quantization parlance, the union of all of the bounded cells S_i will be referred to as the *granular region* of the quantizer and the remaining union of unbounded cells will be referred to as the *overload region* of the quantizer. We will concentrate on the granular region and assume that the overload region contributes a negligible amount to the total average distortion. Before continuing with the development of granular quantization distortion, we make some observations on extending the analysis to include distortion contribution of the overload region.

If Ω is the granular region of the quantizer (the union of all bounded quantization cells), then we can write

$$\begin{aligned}
\bar{d} &= \sum_{S_i \subset \Omega} \int_{S_i} f(x)\rho(||x - y_i||)\,dx + \sum_{S_i \not\subset \Omega} \int_{S_i} f(x)\rho(||x - y_i||)\,dx \\
&= \Pr(X \in \Omega) \sum_{S_i \subset \Omega} \int_{S_i} f_{X|X\in\Omega}(x)\rho(||x - y_i||)\,dx \\
&\quad + \Pr(X \notin \Omega) \sum_{S_i \not\subset \Omega} \int_{S_i} f_{X|X\notin\Omega}(x)\rho(||x - y_i||)\,dx \\
&= \Pr(X \in \Omega)\bar{d}_{\text{granular}} + \Pr(X \notin \Omega)\bar{d}_{\text{overload}},
\end{aligned}$$

where $\bar{d}_{\text{granular}}$ and $\bar{d}_{\text{overload}}$ refer to the conditional average distortion for the granular and overload regions, respectively. This provides the obvious bounds

$$\bar{d}_{\text{granular}} \Pr(X \in \Omega) \leq \bar{d} \leq \bar{d}_{\text{granular}} + \Pr(X \notin \Omega)\bar{d}_{\text{overload}}. \tag{5.2.2}$$

Thus Eq.(5.2.2) provides bounds as well as an approximation that becomes increasingly accurate as $\Pr(X \in \Omega) \to 1$. The first term $\bar{d}_{\text{granular}}$ will be the focus of subsequent analysis, the underlying assumption being that $\Pr(X \notin \Omega)$ is so small that the corresponding contribution to the average distortion is negligible in comparison. Thus the basic approximation is that

$$\bar{d} \approx \bar{d}_{\text{granular}}. \tag{5.2.3}$$

If the quantization reproduction levels are chosen for all cells in an intelligent fashion, e.g., choosing y_i to be the generalized "centroid" of its region S_i, that is,

$$y_i = \min_u E(d(X,u)|X \in S_i)$$

(this is the optimum choice, see [30]), then a simple upper bound to the overload distortion is obtained by mapping all inputs in the overload region into a fixed reproduction level, say the k-dimensional vector 0 (which we will assume to be available). In this case we have the bound

$$\bar{d} \leq \bar{d}_{\text{granular}} + \left(\int_{\Omega^c} f_{X|X \notin \Omega}(x) \rho(||x||)\, dx \right) \Pr(X \notin \Omega),$$

which is often readily computable or boundable.

In practice we can force the importance of granular noise by making the input lie within the granular region before quantizing. In practice this is usually accomplished by "limiting" or "clipping" the input to force it to fall within the granular region. This can be done, for example, using a mapping $g(X_n)$ which leaves the input untouched provided it falls within some allowed region $[-b, b]$ and maps the input into b $(-b)$ if it is greater than b (less than $-b$). The value of b is chosen so that the limited input lies within the granular region. This mapping will incur some distortion (in fact, this is the overload distortion), but then no further overload distortion will occur in the quantization itself. By choosing b sufficiently large, the contribution of the limiting distortion can be made small.

Returning to the general development for granular distortion, we now have the asymptotic approximation

$$\bar{d} \approx \sum_{i=1}^{N} \frac{P_i}{V(S_i)} \int_{S_i} \rho(||x - y_i||)\, dx, \tag{5.2.5}$$

where we have assumed that the overload distortion is negligible.

This high rate approximation has two uses in the analysis of vector quantizers. The first application is a lower bound to the average distortion obtained by considering the optimum shaped cells S_i without regard to whether or not a quantizer can actually have such cells. This bound is an extension of a result of Zador [85] by Yamada, et al. [84] and will provide a lower bound to the performance of high rate vector quantizers of arbitrary structure. We shall see that for a constrained class of norm-based distortion measures, the optimal quantization cell shape (in the sense of minimizing the individual terms of 5.2.5) is a k-dimensional sphere with respect to the seminorm used to define the distortion measure. Unfortunately, one cannot partition k-dimensional space into spheres and hence this bound is overly optimistic.

The second application is an approximation to the actual average distortion (rather than just a lower bound) for a highly structured class of vector quantizers, those whose reproduction levels form a lattice in Euclidean space. Lattice quantizers constitute an important special case of vector quantizers and can be viewed as a multidimensional generalization of the ubiquitous scalar quantizer.

We first focus on the general lower bounds and later return to the case of the lattice quantizers. Both approaches have (5.2.5) as their starting point.

5.3 The High Rate Lower Bound

The following lemma provides a lower bound to each of the terms of the sum by replacing the cells by spheres having the same volume. The result can be viewed as a generalization of the fact that a sphere has the minimum possible moment of inertia with respect to its centroid of any convex polytope with the same volume. To state the lemma we require a definition. Given a bounded set S with volume $V(S)$, define the *equivalent spherical region $T_y(S)$ centered at y* as the k-dimensional sphere centered at y having the same volume as S. Thus

$$T_y(S) = \{x : \; ||x - y|| \leq R(S)\}$$

where

$$R(S) = \left(\frac{V(S)}{V_k}\right)^{1/k}$$

is called the *effective radius* of S. Recall that V_k is the volume of the unit sphere with respect to the given seminorm in k-dimensional space. To explain the above formula and the name of $R(S)$ observe that the volume of a sphere of radius a centered at y is given by

$$V(x : \; ||x - y|| \leq a) = V(x : ||x|| \leq a) = \int_{x:\; ||x|| \leq a} dx$$

$$= \int_{x:||x/a|| \leq 1} dx = \int_{u:||u|| \leq 1} a^k \, du = a^k V_k,$$

where we have used the fact that volume is invariant to translation. Thus $R(S)$ is the radius of a sphere which has the same volume as S as claimed.

Lemma 5.3.1 *Given a seminorm-based distortion measure $d(x, y) = \rho(||x - y||)$, then for any bounded set*

$$\int_S \rho(||x - y||) \, dx \geq \int_{T_y(S)} \rho(||x - y||) \, dx.$$

In words, the average distortion resulting by representing a vector with a uniform distribution in a set S by a single point y is bound below by the average distortion resulting when the set S is replaced by a sphere of the same volume.

Proof:

Since y is fixed, abbreviate $T_y(S)$ to $T(S)$. The set $S \bigcup T(S)$ can be written as $S \bigcup (T(S) \bigcap S^c)$ and $T(S) \bigcup (S \bigcap T(S)^c)$ and hence

$$\int_{S \bigcup (T(S) \bigcap S^c)} \rho(||x-y||)\,dx = \int_{T(S) \bigcup (S \bigcap T(S)^c)} \rho(||x-y||)\,dx$$

which implies that

$$\int_S \rho(||x-y||)\,dx + \int_{T(S) \bigcap S^c} \rho(||x-y||)\,dx$$

$$= \int_{T(S)} \rho(||x-y||)\,dx + \int_{S \bigcap T(S)^c} \rho(||x-y||)\,dx$$

so that

$$\int_S \rho(||x-y||)\,dx = \int_{T(S)} \rho(||x-y||)\,dx +$$

$$\int_{S \bigcap T(S)^c} \rho(||x-y||)\,dx - \int_{T(S) \bigcap S^c} \rho(||x-y||)\,dx.$$

The lemma will be proved by the above formula if we can show that

$$\int_{S \bigcap T(S)^c} \rho(||x-y||)\,dx - \int_{T(S) \bigcap S^c} \rho(||x-y||)\,dx \geq 0.$$

Recall that $x \in T(S)$ if and only if $||x-y|| \leq R(S)$. Since we are considering a seminorm-based distortion measure, this is true if and only if $d(x,y) = \rho(||x-y||) \leq \rho(R(S))$. Thus the integrand of the left hand integral above is bound below by $\rho(R(S))$ and the integrand in the right hand integral is bound above by $\rho(R(S))$ and hence the difference is bound below by

$$\rho(R(S)) \int_{S \bigcap T(S)^c} dx - \rho(R(S)) \int_{T(S) \bigcap S^c} dx$$

$$= \rho(R(S))(V(S \bigcap T(S)^c) - V(T(S) \bigcap S^c)).$$

We will be done if we can show that the difference of the volumes inside the parentheses is nonnegative. Since $S = (S \bigcap T(S)^c) \bigcup (S \bigcap T(S))$ and hence $V(S) = V(S \bigcap T(S)^c) + V(S \bigcap T(S))$ since the union is of disjoint sets, we have that

$$V(S \bigcap T(S)^c) - V(T(S) \bigcap S^c) = V(S) - V(S \bigcap T(S)) - V(T(S) \bigcap S^c)$$

$$= V(S) - V(T(S)),$$

where we have used a similar argument to conclude that $V(S \bigcap T(S)) + V(S^c \bigcap T(S)) = V(T(S))$. By construction of $T(S)$, however, $V(S) = V(T(S))$, which completes the proof.

The above lemma can be combined with (5.2.5) to provide a lower bound to the average distortion. Before writing the lower bound, however, we further develop the average distortion over spherical regions. We have that

$$\int_{T(S)} \rho(||x-y||)\, dx = \int_{x:\, ||x-y|| \leq R(S)} \rho(||x-y||)\, dx = \int_{x:\, ||x|| \leq R(S)} \rho(||x||)\, dx$$

$$= \int_{x:\, ||x/R(S)|| \leq 1} \rho(||x||)\, dx = R(S)^k \int_{u:\, ||u|| \leq 1} \rho(R(S)||u||)\, du,$$

where $u = x/R(S)$. Using the fact that $R(S) = (V(S)/V_k)^{1/k}$ by definition, we have that

$$\int_{T(S)} \rho(||x-y||)\, dx = \frac{V(S)}{V_k} \int_{u:||u|| \leq 1} \rho((\frac{V(S)}{V_k})^{1/k} ||u||)\, du.$$

Note in particular that the right-hand side does not depend on y.

Define the *Gish-Pierce function* $M_k(v)$ by

$$M_k(v) = \frac{1}{V_k} \int_{u:||u|| \leq 1} \rho(v V_k^{-1/k} ||u||)\, du.$$

Then we can write

$$\int_{T(S)} \rho(||x-y||)\, dx = V(S) M_k(V(S)^{1/k}).$$

$M_k(v)$ can be thought of as the average distortion resulting when a random vector with a uniform distribution on a sphere of volume v^k is represented by a point at the center of the sphere.

Combining the above formula with (5.2.5) we have the following lower bound to the average distortion:

$$\bar{d} \geq \sum_{i=1}^{N} P_i M_k(V(S_i)^{1/k}).$$

We now proceed to approximate this sum by an integral. In preparation, we introduce the idea of a reproduction vector density $g_N(x)$ defined by

$$g_N(x) = \frac{1}{N V(S_i)}, \quad \text{if } x \in S_i,\ i = 1, 2, \ldots, N.$$

Suppose that as $N \to \infty$ this quantity has a limiting value $\lambda(x)$ which has unit integral. Then, for sufficiently large N, $\lambda(x)\Delta V(x)$ is the fraction of reproduction vectors in an incremental volume containing x and the total number of reproduction vectors in a set is N times the integral of the density over the set. Thus for large N, $g_N(x)$ approximates its limit and $V(S_i)$ is therefore approximately $1/(N\lambda(y_i))$. This gives us the approximate lower bound

$$\bar{d} \geq \sum_{i=1}^{N} P_i M_k((N\lambda(y_i))^{-1/k}).$$

Recall that $f(x)$ is approximately $P_i/V(S_i)$ for any $x \in S_i$. Because the volumes are small and the reproduction vectors dense, y_i must be in S_i or very close to it and hence $f(y_i)$ is also approximately $P_i/V(S_i)$ and hence

$$\bar{d} \geq \sum_{i=1}^{N} f(y_i)V(S_i)M_k((N\lambda(y_i))^{-1/k}).$$

As N becomes large the volumes $V(S_i)$ become small and approximate a volume differential and hence the above sum can be approximated by the integral

$$\bar{d} \geq \int dy f(y) M_k((N\lambda(y))^{-1/k}) = EM_k((N\lambda(X))^{-1/k}). \tag{5.3.1}$$

We define the final quantity above as the *high rate quantizer distortion lower bound* and we denote it by $\bar{d}_L$.

The bound can be put in a more useful form by evaluating the integral defining M_k. For the assumed case of a seminorm-based distortion measure we have that

$$M_k(v) = \frac{1}{V_k} \int_{u:||u|| \leq 1} \rho(V_k^{-1/k} v ||u||)\, du.$$

The above integral can be evaluated as (see [84])

$$M_k(v) = \int_0^1 \rho((\alpha/V_k)^{1/k} v)\, d\alpha = k \int_0^1 \rho(V_k^{-1/k} v\beta)\beta^{k-1}\, d\beta.$$

For the particular form $\rho(\alpha) = \alpha^r$, $r \geq 1$, this becomes

$$M_k(v) = v^r V_k^{-r/k} \int_0^1 \alpha^{r/k}\, d\alpha = v^r \frac{k}{k+r} V_k^{-r/k}.$$

Note that the above integrals are one-dimensional integrals.

We now summarize the main results of this section: Given a k dimensional quantizer with a seminorm-based distortion measure $d(x, y) =$

$||x-y||^r$, where $||\cdot||$ is a seminorm in k-dimensional Euclidean space and $r \geq 1$, then

$$\bar{d} = Ed(X, q(X)) \geq \bar{d}_L = EM_k((N\lambda(X))^{-1/k})$$
$$= \frac{k}{k+r}(NV_k)^{-r/k}E(\lambda(X)^{-r/k}), \tag{5.3.2}$$

where $\lambda(x)$ is the asymptotic quantizer density. Later we will obtain bounds on the optimal performance by optimizing over λ subject to constraints on the coding rate.

As a simple example, note that if $\lambda(x)$ is chosen to be a constant over some bounded region S with volume V, then since $\lambda(x)$ must have unit integral,

$$\lambda(x) = \frac{1}{V}1_S(x) \tag{5.3.3}$$

and hence

$$\bar{d}_L = \frac{k}{k+r}(N\frac{V_k}{V})^{-r/k}.$$

If $\lambda(x)$ is constant over a region, then the quantizer output vectors are uniformly distributed through this region. For example, they could be the points corresponding to an ordinary scalar uniform quantizer used k times to quantize a vector. The points formed in this way form an example of a k-dimensional rectangular lattice. We shall shortly define lattices more carefully as they provide a highly structured class of vector quantizers with several nice properties.

If the density $\lambda(x)$ of quantization intervals is constant, then one can think of the quantization levels being spread uniformly throughout some region of space so that the corresponding Voronoi regions are all congruent, that is, if the points are uniformly spread, the collection of nearest neighbors around each point should look the same for most points (except those near the boundary of the granular region). In particular, these cells should all have the same volume, which we define as

$$\Delta = \frac{V}{N};$$

that is, the volume of the granular region divided by the total number of cells. With this definition we have that

$$\bar{d}_L = \frac{k}{k+r}(\frac{V_k}{\Delta})^{-r/k}. \tag{5.3.4}$$

If the squared error distortion is used, the $r=2$ and V_k can be applied from Eq. (4.8.3) to obtain

$$\bar{d}_L = \frac{k}{k+2}(\frac{2}{\Delta k\Gamma(k/2)})^{-2/k}. \tag{5.3.5}$$

5.4 High Rate Entropy

In a manner similar to that used to approximate the average distortion of a vector quantizer when the rate is high, we can approximate the entropy of the quantizer output. This is primarily useful if subsequent noiseless coding is to be permitted, in which case the entropy of the quantized signal is the appropriate measure of coding rate. (See any book on information theory or Storer [77] or Gersho and Gray [30] for treatments of noiseless coding.)

The entropy of the quantized vector is given by

$$H = H(q(X)) = -\sum_{i=1}^{N} P_i \log P_i,$$

where $P_i = \int_{S_i} f(x)\, dx$. As before we make the approximation that P_i is nearly $f(y_i)/(N\lambda(y_i)) = f(y_i)V(S_i)$ and therefore that

$$H \approx -\sum_{i=1}^{N} \frac{f(y_i)}{N\lambda(y_i)} \log \frac{f(y_i)}{N\lambda(y_i)}$$

$$= -\sum_{i=1}^{N} V(S_i) f(y_i) \log f(y_i) + \sum_{i=1}^{N} V(S_i) f(y_i) \log(N\lambda(y_i))$$

which as in the previous section we can approximate by the integral

$$-\int dy f(y) \log f(y) + \int dy f(y) \log(N\lambda(y)) \approx h(X) - E(\log \frac{1}{N\lambda(y)}),$$

where $h(X)$ is the differential entropy of the random vector X. We therefore have the approximation for large N that

$$H(q(X)) \approx h(X) - E(\log \frac{1}{N\lambda(y)}). \tag{5.4.1}$$

This formula is of independent interest because it relates the actual entropy of the quantized vectors to the differential entropy of the input. It thereby provides a connection between differential entropy and entropy.

As a simple, but important, example, again consider the case of a vector uniform quantizer with λ given by (5.3.3). In this case,

$$H(q(X)) \approx h(X) - \log \frac{V}{N} = h(X) - \log \Delta, \tag{5.4.2}$$

where again we have defined Δ as the volume of a single Voronoi cell.

5.5 Lattice Vector Quantizers

We have observed that both the average distortion and entropy have very simple high rate approximations when the quantization level density $\lambda(x)$ is uniform. In this section we define a class of quantizers which have a constant level density, the *lattice quantizers* whose reproduction points form a regular lattice in k dimensional space. We shall see that for lattices it is easy to find (or at least write down) a good high rate approximation for the average distortion. This approximation can then be compared to the previously derived lower bound.

Our treatment of lattice quantizers will be quite brief, we will not treat their principal attraction: many fast and simple algorithms exist for actually implementing lattice vector quantizers and the performance improvement over scalar quantizers is worth the effort. The interested reader is referred to Conway and Sloane [22] and Gibson and Sayood [31] for thorough treatments.

Suppose now that the collection $\{y_i;\ i = 1,\ldots,N\}$ is chosen from a subset of a lattice in $\mathcal{R}^k$. A k-dimensional *lattice* Λ in $\mathcal{R}^k$ has the form $\Lambda = \{\sum_{i=0}^{k-1} l_i\nu_i;\ \text{all integers } l_i;\ i = 0,1,\ldots,k-1\}$, where $\nu_0,\ldots,\nu_{k-1}$ are k linearly independent vectors in $\mathcal{R}^k$ called the *generator* vectors for the lattice. Thus a lattice consists of all integral linear combinations of its generator vectors. The lattice goes on forever, but any quantizer uses only a finite subset of the lattice for reproduction points. The choice of *which* finite collection of points is important in practice, but here we will continue to focus on the quantizer behavior only in the region where the reproduction points are and we ignore the effects of overload, effectively assuming that the total distortion contribution due to overload is negligible with respect to that within the region covered by the points.

A lattice automatically partitions $\mathcal{R}^k$ into the collection of nearest-neighbor or *Voronoi* regions corresponding to the lattice points. The Voronoi region of a lattice point contains all of the points in $\mathcal{R}^k$ closer to that point than to any other point. (As before, we assume that there is negligible probability on the boundaries between the Voronoi regions.) Suppose that S_0 is the Voronoi region of the lattice point 0 (k-dimensional 0), which we assume to be a quantizer reproduction vector. Then the Voronoi region of any lattice point a will simply be $a + S_0$, that is, a region of the same shape and size as S_0, but "centered" at a instead of 0. Because of this fact we have using a simple variable change that for any difference distortion measure

$$\int_{S_i} \rho(||x - y_i||)\,dx = \int_{S_0} \rho(||x||)\,dx,$$

a quantity independent of the particular reproduction level y_i or its index

i. Furthermore, all of the Voronoi regions (ignoring the overload regions) have the same volume so that (5.2.5) becomes

$$\bar{d} \approx \sum_{i=1}^{N} \frac{P_i}{V(S)} \int_{S_i} \rho(||x - y_i||)\, dx = \frac{1}{V(S_0)} \int_{S_0} \rho(||x||)\, dx. \qquad (5.5.1)$$

For example, for the case of the Euclidean norm and $\rho(u) = u^2$, we have that

$$\bar{d} \approx \frac{1}{V(S_0)} \int_{S_0} ||x||_2^2\, dx. \qquad (5.5.2)$$

This normalized second moment (and various moments related to it) has been evaluated for a variety of lattices [22]. As a simple example, suppose that we consider a rectangular lattice with orthogonal vectors of length b as generators and we consider the squared error distortion measure. Then the volume of a Voronoi cell is b^k and the average distortion normalized to the number of dimensions is given by

$$\frac{1}{k}\bar{d} = \frac{1}{k} \frac{\int_{-b/2}^{b/2} \cdots \int_{-b/2}^{b/2} \sum_{j=0}^{k-1} x_i^2\, dx_0 \ldots dx_{k-1}}{b^3} = \frac{b^2}{12}, \qquad (5.5.3)$$

an ubiquitous formula in the quantization literature.

This formula provides a relatively simple means of evaluating the average distortion of a high rate lattice vector quantizer with a smooth input probability. No optimization has been done, so that one would expect the distortion to be larger than that predicted by the previous sections and larger than that predicted using the equivalent sphere argument. On the other hand, many good lattices are known which have fast and simple encoding algorithms. The above formula shows that knowing only a form of generalized moment, one has an estimate of the average distortion for the entire quantizer. (See, e.g., [22].) The lattices minimizing (5.5.2) for dimensions 1 and 2 are the uniform lattice and the hexagonal lattice, respectively. A discussion of the best known lattices for higher dimensions may be found in [22].

5.6 Optimal Performance

We next develop lower bounds to the optimal performance of high rate vector quantizers. These bounds serve the same purpose that the rate distortion or distortion rate functions serve in combination with the negative source coding theorem. The derivation here is accomplished by the straightforward application of standard integration inequalities, Jensen's and Hölder's inequalities[46], in particular, to the previously derived approximations.

First suppose that the number of quantization levels N is fixed. Hölder's inequality states that given two integrable functions $h(x)$ and $g(x)$ and two positive real numbers p and q with the property that $1/p + 1/q = 1$, then

$$\int |h(x)g(x)|\, dx \leq ||h||_p ||g||_q,$$

where here the norm $||\cdot||$ is the L_p norm defined by

$$||h||_p = (\int |h(x)|^p\, dx)^{1/p}.$$

Equality holds in the Hölder inequality if and only if

$$\left(\frac{|h(x)|}{||h||_p}\right)^p = \left(\frac{|g(x)|}{||g||_q}\right)^q.$$

We can apply Hölder's inequality to obtain a lower bound to $\bar{d}_L$ by setting $p = (k+r)/k$, $q = kp/r = (k+r)/r$, $h^p = f\lambda^{-r/k}$, and $g^q = \lambda$ to obtain

$$E(\lambda(X)^{-r/k}) = \int f(x)\lambda(x)^{-r/k}\, dx \geq ||f||_{k/(k+r)}$$

with equality if and only if $\lambda(x)$ is proportional to $f(x)^{k/(k+r)}$. This provides the general asymptotic quantizer lower bound

$$\bar{d} \geq \bar{d}_L \geq \frac{k}{k+r}(V_k N)^{-r/k} ||f||_{k/(k+r)}. \quad (5.6.1)$$

The rightmost equality is achieved if and only if

$$\lambda(x) = \frac{f(x)^{k/(k+r)}}{||f||_{k/(k+r)}^{k/(k+r)}}.$$

Note that the only dependence of the bound on the particular seminorm used to define the distortion measure is through the volume of the unit sphere with respect to that seminorm, V_k. Thus the previously stated values for V_k for particular seminorms coupled with the above formula provide several examples of asymptotic quantizer lower bounds.

The previous bound provides a lower bound to the average distortion for a fixed (large) value of N and hence a fixed value of the rate $\log N$. Alternatively, the entropy of the output vector can be used as a measure of rate, e.g., if subsequent noiseless coding is to be used. We next consider a bound explicitly involving entropy.

Jensen's inequality states that if ϕ is a convex $\bigcup$ function, then

$$\phi(EX) \leq E\phi(X)$$

with equality if and only if X has a uniform distribution over some set. Since $-\log$ is a convex function, we have that

$$H \approx h(X) - kE(\log((\frac{1}{N\lambda(X)})^{1/k}) \geq h(X) - k \log E((\frac{1}{N\lambda(X)})^{1/k}).$$

Thus we have approximately that

$$\log E((\frac{1}{N\lambda(X)})^{1/k}) \geq \frac{1}{k}(h(X) - H(q(X)))$$

or

$$E((\frac{1}{N\lambda(X)})^{1/k}) \geq e^{-\frac{1}{k}(H(q(X))-h(X))}. \tag{5.6.2}$$

Similarly, since α^r is a convex function for $r \geq 1$, we can apply Jensen's inequality to (5.3.2) to obtain

$$\bar{d}_L \approx \frac{k}{k+r}(V_k)^{-r/k}E((\frac{1}{N\lambda(X)})^{r/k}) \geq \frac{k}{k+r}(V_k)^{-r/k}(E(\frac{1}{N\lambda(X)})^{1/k})^r.$$

Combining this bound with (5.6.2) then yields

$$\bar{d}_L \geq \frac{k}{k+r}(V_k)^{-r/k}e^{-\frac{r}{k}(H(q(X))-h(X))}, \tag{5.6.3}$$

with equality if and only if $\lambda(x)$ is a constant, that is, the quantizer reproduction vectors are uniformly distributed over some set having probability 1. This bound is called the *constrained entropy high rate quantizer bound*.

Observe that the bound is achieved by high rate lattice vector quantizers since they have a uniform density of quantization levels. Thus lattice vector quantizers are optimal in a sense when rate is measured by the entropy of the quantized vector rather than by the log of the number of quantizer outputs.

5.7 Comparison of the Bounds

In this section we make some simple comparison of the bounds and make more explicit their dependence on the dimension k. We can summarize the two bounds making the dimension k explicit as

$$\bar{d}_L(k) \geq \frac{k}{k+r}(V_kN)^{-r/k}||f_k||_{k/(k+r)} \tag{5.7.1}$$

for a constrained number N of quantization levels and

$$\bar{d}_L(k) \geq \frac{k}{k+r}(V_k)^{-r/k} e^{-\frac{r}{k}(H(q(X^k))-h(X^k))}, \tag{5.7.2}$$

for a constrained quantizer entropy. Note that the bounds are the same except for the $N^{-r/k}||f||_{k/(k+r)}$ term in the first being replaced by the $e^{-\frac{r}{k}(H(q(X))-h(X))}$ term in the second. Note also that different quantizer densities achieve the two bounds.

In the first bound we are constraining the number of quantizer reproduction vectors and hence we are constraining the rate defined as $R = k^{-1}\log_2 N$ in bits per symbol. Hence we can define the constrained rate high rate quantizer bound by

$$\bar{d}_k^{\text{rate}}(R) = \frac{k}{k+r} V_k^{-r/k} e^{-rR} ||f_k||_{k/(k+r)}. \tag{5.7.3}$$

Similarly, in the second bound we are constraining the entropy and hence we are effectively constraining a rate defined as $R = k^{-1}H(q(X^k))$ (which gives the average codeword length achievable using noiseless coding and hence the average rate). Thus we can define the constrained entropy high rate quantizer bound by

$$\bar{d}_k^{\text{entropy}}(R) = \frac{k}{k+r}(V_k)^{-r/k} e^{-r(R-\frac{1}{k}h(X^k))}. \tag{5.7.4}$$

Intuitively one would suspect that for a fixed R the average distortion achievable using constrained entropy would be lower than that using constrained code rate since if the code rate is smaller than R, so is the entropy, that is, constraining the code rate to be less than R is a stronger constraint. We can easily verify this observation by using Jensen's inequality to obtain the following inequalities:

$$\log ||f_k||_{k/(k+r)} = \frac{k+r}{k} \log \int dx f_k(x) e^{-\frac{r}{k+r}\log f_k(x)}$$

$$\geq \frac{k+r}{k} \int dx f_k(x)\left(-\frac{r}{k+r}\log f_k(x)\right) = -\frac{r}{k}\int f_k(x)\log f_k(x)\, dx$$

whence

$$||f_k||_{k/(k+r)} \geq e^{\frac{r}{k}h(X^k)}.$$

This implies that

$$\bar{d}_k^{\text{rate}}(R) \geq \bar{d}_k^{\text{entropy}}(R) \tag{5.7.5}$$

as claimed.

It is also of interest to compare the asymptotic quantization bounds with the Shannon lower bounds of the previous chapter. We can restate the Shannon lower bound for the rth power of a seminorm using (4.8.9) as

$$D_k(R) \geq (er(V_k\Gamma(1+\frac{k}{r}))^{r/k})^{-1}e^{-r(R-h_k)} = D^k_{\mathrm{SLB}}(R) \qquad (5.7.6)$$

where

$$h_k = \frac{1}{k}h(p_k).$$

The Shannon lower bound can be compared to the asymptotic quantization lower bound of (5.6.3) by observing that $H(q(X)) \leq R$ and hence the average distortion can be bound below by

$$\frac{1}{k}\bar{d}_L = \frac{1}{k}\bar{d}_k^{\mathrm{entropy}} = \frac{V_k^{-r/k}}{r+k}e^{-r(R-h_k)} = \left(\frac{e\Gamma(1+\frac{k}{r})^{r/k}}{1+\frac{k}{r}}\right) D^k_{\mathrm{SLB}}(R). \qquad (5.7.7)$$

It can be shown that the factor multiplying the Shannon lower bound to get the asymptotic quantizer lower bound is no smaller than 1 and hence the asymptotic quantizer bounds are better (larger) lower bounds (as should be expected)[84]. It can further be shown using Stirling's approximation that this factor tends to one as $k \to \infty$, that is, the two bounds coincide in the limit as $k \to \infty$, which also should be expected since as $k \to \infty$ optimal vector quantizers should perform near the Shannon limit because of the source coding theorem.

5.8 Optimized VQ vs. Uniform Quantization

The comparisons among the various bounds has a fascinating and important implication regarding the choice of vector quantization technique. We have seen that if coding rate (log the number of reproduction levels) is the essential constraint, then one can in theory choose the vector quantizer (or VQ) reproduction levels in an optimum fashion and approximately achieve certain lower bounds. Actually designing such optimum quantizers can be a difficult task, although a variety of design algorithms for good codes exist [30]. On the other hand, if one instead uses entropy to measure rate, which implicitly means that one will use additional complexity to noiselessly code the quantized vectors so that the actual number of transmitted bits is close to the entropy, then the optimum vector quantizer should have a uniform density of reproduction levels in k-dimensional space, e.g., it can be a lattice vector quantizer in general or a simple scalar uniform quantizer in particular. Since uniform quantizers are simple to build and use, the trade-off here seems to be between having a complicated VQ with no noiseless coding or having a simple VQ with possibly complicated noiseless coding. The question now is this: We have seen that optimized VQs can perform arbitrarily close to the Shannon limit given by the rate-distortion function. Can uniform scalar quantizers combined with noiseless codes also get arbitrarily close to the Shannon limit? If not, can they get close enough so that the potential reduction in complexity using such "simple" codes more than offsets the loss of performance? In this section we focus the results of the previous section to compare the performance of uniform quantizers to the optimal performance promised by Shannon theory for a squared error distortion measure and then discuss the implications of the results.

From (5.5.3) we have that the average distortion per symbol resulting from using a high rate uniform quantizer with reproduction levels spaced by b is

$$\bar{d} = E[(X_0 - q(X_0))^2] \approx \frac{b^2}{12},$$

where we have invoked the stationarity of the source and the fact that we are using a scalar quantizer k times to form our k-dimensional vector quantizer. Define $D = \bar{d}$ to ease comparison with the rate-distortion bounds. For any k, the kth order entropy of the quantized signal is given by

$$H_k \approx kh_k - \log(b^k)$$

or

$$\frac{1}{k}H_k = h_k - \log b.$$

Taking the limit as the dimension $k \to \infty$, the left hand side gives the entropy rate of uniformly quantized process, the right hand side is the differential entropy rate of the source minus $\log b$, where b is the distance between the levels of the uniform quantizer, that is,

$$\bar{H}(q) \approx \bar{h} - \log b.$$

From the average distortion approximation, however, $b \approx \sqrt{12D}$ and hence

$$\bar{H}(q) \approx -\frac{1}{2}\log D - \frac{1}{2}\log 12 + \bar{h}.$$

But the Shannon lower bound to the rate-distortion function is given by (4.8.12) as

$$R_{\mathrm{SLB}}(D) = \bar{h} - \frac{1}{2}\log(2\pi e D).$$

Combining all of these facts we have that for high rate uniform scalar quantization,

$$\bar{H}(q) - R_{\mathrm{SLB}}(D) \approx -\frac{1}{2}\log(12D) + \frac{1}{2}\log(2\pi e D) = \frac{1}{2}\log(\frac{\pi e}{6}) \approx .2546.$$

Since the Shannon lower bound is smaller than the true rate-distortion function, if we assume that the approximate equality is very close to equality we have shown that

$$\bar{H}(q) \leq R(D) + .255. \tag{5.8.1}$$

On its face, this is a discouraging bound since it says that uniform quantization combined with noiseless coding performs almost within one quarter of a bit of the Shannon optimum. (Even though the Shannon optimum was proved for fixed rate codes, it can be extended to variable-rate codes as well. See, for example, [54].)) This suggests that indeed one might do well to not try to design clever vector quantizers and instead be content with simple uniform quantizers and noiseless codes. This discouraging result was first developed for memoryless sources by Goblick and Holsinger [33] and extensions are discussed, e.g., in Ziv [88] and Neuhoff [67].

The result certainly provides theoretical support for always considering the uniform quantizer/noiseless coding alternative. It does not, however, render optimal quantization uninteresting for several reasons. The primary reason is complexity, as pursued in some detail by Neuhoff [67]. Noiseless codes have reasonable complexity when run on sources with moderate alphabet size. To achieve the above bound, however, they must yield average codeword lengths near the entropy rate of the quantized sequence. This means that one must noiselessly encode vectors from the uniform quantizers. Such vectors can have large alphabets so that the noiseless codes become quite complicated. One may be able to trade off the complexity between the vector quantizer and noiseless code, but one cannot avoid the complexity entirely: Near optimal performance with a uniform quantizer can involve using quite complex noiseless codes.

Other reasons make optimal quantization important in some applications. The development assumed high rate quantizers, but many important applications involve low rates, that is, trying to communicate data such as speech or images at fractions of a bit per sample. Most sophisticated source coding algorithms are aimed at low bit rates and most high rate systems indeed use either uniform quantization alone or in tandem with a simple nonlinear memoryless mapping (compandor). Ziv [88] has shown that even in the low rate case, uniform quantization combined with noiseless coding can operate within .754 bits of the rate-distortion function. If the rate is already only .5 bits per sample or less (as is often the case with images and speech), one cannot conclude that the uniform quantizer is good enough to not bother looking at optimized quantizers.

By operating on vectors one can use more interesting distortion measures better matched to the nature of the particular source. An example is the Itakura-Saito distortion measure in speech coding systems. Such systems can achieve much better quality at low rates than can any squared error based waveform coding systems[30].

Lastly, optimized quantizers may have structural properties lacking in the uniform quantizer/noiseless code cascade that are useful in some applications. For example, tree-structured vector quantizers have a successive approximation property that means quality improves with the arrival of additional bits. This makes such tree structured codes especially amenable to progressive transmission and variable-rate environments[30].

5.9 Quantization Noise

An interpretation of the high rate development leads to an intuitive description of quantization noise which is commonly used as the basic approximation of quantization noise analysis, sometimes with unfortunate results. The fundamental assumption of high rate quantization analysis was that the probability density function of the input vector x was approximately constant over each (small) Voronoi region of the quantizer. Thus *given* that $x \in S_i$, the conditional density function is approximately

$$f_{X|X\in S_i}(x) \approx \frac{1}{V(S_i)} \text{ for } x \in S_i.$$

Consider the quantization noise or error

$$\epsilon = \epsilon(x) = q(x) - x.$$

The choice of sign is to yield the formula

$$q(x) = x + \epsilon \tag{5.9.1}$$

representing the quantizer output as the sum of the input and a noise term (the so-called *additive noise model* for quantization error). We will refer to both ϵ and $-\epsilon$ as the "quantization noise," as convenient. If the input vector pdf is approximately uniform over a set S_i, then the quantizer error vector, $-\epsilon = x - q(x)$, will also be uniform, but it will be shifted by the reproduction level y_i. In other words, it will be distributed like $x - y_i$ with x uniform on S_i or, equivalently, it will be distributed uniformly on the set $S_i - y_i$. Even in the high rate case, the error will have clearly different behavior depending on the particular quantization cell. It may always be uniformly distributed, but the shape of the region over which it is uniform will differ. Thus, for example, the maximum error will vary from cell to cell. Note that the error distribution will be exactly uniform if the input vector distribution was itself exactly uniform over the granular region of the quantizer.

If the quantizer is a lattice quantizer, however, then all of the Voronoi regions will have the same shape and volume and the error will be uniformly distributed over S_0 *regardless* of the particular quantization level. From (5.9.1) we can then consider the quantizer output to be formed by adding the input vector to a random vector having approximately a uniform distribution over a specific finite volume set S_0, the Voronoi region at the origin. For this reason, ϵ is often called "quantization noise" or "quantizer noise." More accurately it should be considered as the granular quantization noise since we have assumed that there is no overloading. The reader should beware, however, as ϵ is not really noise in the usual sense: it is a deterministic function of the input and hence cannot be statistically independent of the input. The high rate approximation, however, can be interpreted as an assumption that the quantizer noise vector is approximately uniformly spread over the Voronoi regions of the quantizer outputs. As we have seen, this can significantly simplify the mathematics.

Suppose that the lattice Λ is a *rectangular* lattice, e.g., the generator vectors are orthogonal (say unit vectors in k-dimensional Euclidean space). In this case ϵ is approximately uniform over the Voronoi region of 0, which is just a k-dimensional cube. This suggests that not only is the noise pdf uniform, it can be considered as a product of k 1-dimensional uniform pdfs; that is, the quantization noise looks approximately like a k-dimensional uniformly distributed independent identically distributed random sequence, a form of white noise!

The white noise approximation for quantization noise is originally due to Bennett [4] and it has become the most common approximation in the analysis of quantization systems. Note that the above derivation of this approximation really only makes sense if

1. The quantization is uniform.

2. The quantizer does not overload.

3. The rate is high.

4. The maximum size of the quantizer cells is small.

5. The input joint density is smooth.

Unfortunately, the assumption is often made even when these conditions do not hold. In the final chapter we explore in some detail the example of a uniform quantizer and find an exact analysis for the quantization noise which can be compared to the high rate approximations.

Exercises

1. Evaluate the high rate quantizer bounds to optimal performance for both the entropy and rate constraints for a fixed k dimensional random vector X for the cases given below. Compare the result to the k-dimensional vector Shannon lower bound to the rate distortion function .
   (a) iid Gaussian, squared error distortion measure.
   (b) Uniform distribution over a k-dimensional sphere of radius b and a magnitude error distortion measure.
   (c) Uniform distribution over a k-dimensional cube with side b and a magnitude error distortion measure.
   (d) Gauss Markov source with correlation coefficient r, variance σ^2, and squared error distortion measure.
2. What are the optimum quantizer level densities $\lambda(x)$ in the previous problem? What are the resulting quantizer output entropies?
3. Where possible, evaluate the various bounds in the first problem for the case $k \to \infty$.
4. Repeat the first problem for the special case of a memoryless Laplacian source and a magnitude error distortion measure.
5. Prove that
$$\left(\frac{e\Gamma(1+\frac{k}{r})^{r/k}}{1+\frac{k}{r}}\right) \geq 1.$$
   Use Stirling's approximation to prove that this quantity goes to 1 in the limit as $k \to \infty$. Thus the statements following (5.7.7) are justified.
6. Compare the high rate approximation of (5.5.2) with the high rate lower bound of (5.3.5). How different is the distortion in dB for a fixed rate? How different are the rates in bits for a fixed distortion?

Chapter 6

Uniform Quantization Noise

6.1 Introduction

We now turn to a theoretical study of the most important practical source coding system: uniform quantization. This example provides a guinea pig for several of the theoretical bounds and approximations of this book. We shall see that some of the aysmptotic theory provides useful insight into the quality of simple scalar quantization, but that other aspects of the theory can be extremely misleading if they are applied carelessly to systems violating the basic conditions required for the approximation arguments to hold.

Uniform quantization has long been a practical workhorse for analog-to-digital conversion (ADC) as well as a topic of extensive theoretical study. Deceptively simple in its description and construction, the uniform quantizer has proved to be surprisingly difficult to analyze precisely because of its inherent nonlinearity. The difficulty posed by the nonlinearity usually has been avoided by simplifying (linearizing) approximations or by simulations. Approximating the quantizer noise in system analysis by uniformly distributed white noise is extremely popular, but it can lead to inaccurate predictions of system behavior. These problems have been known since the earliest theoretical work on quantization noise spectra, but the modern literature often overlooks early results giving conditions under which the approximations are good and providing exact techniques of analysis for certain input signals where the approximations are invalid.

Many of the original results and insights into the behavior of quantization noise are due to Bennett [4] and much of the work since then has its origins in that classic paper. Bennett first developed conditions under which quantization noise could be reasonably modeled as additive white noise. Less well known are the 1947 papers by Clavier, Panter, and Grieg [20], [21] which provided an exact analysis of the quantizer noise resulting when a uniform quantizer is driven by one or two sinusoids. The quantization noise in this case proves to be quite nonwhite in behavior.

By far the most common assumption in the communications literature is that partially described at the end of the previous chapter: the quantization noise is assumed to be a signal-independent, uniformly distributed, uncorrelated (white) additive noise source. This approximation linearizes the system and allows the computation of overall performance and signal-to-quantization noise ratios (SQNR) using ordinary linear systems techniques. Part of the model is obviously wrong since the quantization noise is in fact a deterministic function of the input signal (and hence cannot be signal independent), yet in many applications these assumptions or approximiations yield reasonable results. Among other things, Bennett proved that under certain conditions the white noise assumption provides a good approximation to reality. As developed during the discussion of high rate quantization theory, these assumptions are that

(1) the quantizer does not overload,

(2) the quantizer has a large number of levels,

(3) the bin width or distance between the levels is small, and

(4) the probability distribution of pairs of input samples is given by a smooth probability density function.

If these conditions hold, then the white noise approximation yields good results in a precise sense.

The white noise approximation gained a wide popularity, largely due to the work of Widrow [80]. Unfortunately, the conditions under which the white noise approximation is valid are often violated, especially in feedback quantizers such as Delta modulators and Sigma-Delta modulators, systems in common use in oversampled analog-to-digital converters. The quantizers in these systems typically have only a few levels (often two) that are not close to each other and the feedback of a digital signal prevents the quantizer input from having a smooth joint distribution. Not surprisingly, it was found in some oversampled ADCs (such as the simple single loop Sigma-Delta modulator) that the quantizer noise was not at all white and that the noise contained discrete spikes whose amplitude and frequency depended on the input [15]. Perhaps surprisingly, however, simulations and actual circuits of higher order Sigma-Delta modulators and of interpolative coders often exhibited quantizer noise that appeared to be approximately white. Unfortunately, these systems also often exhibited unstable behavior not predicted by the white noise analysis.

White quantization noise is not an inherently important mathematical property for a system (although it can be subjectively important in speech and image digitization for smoothing out quantization effects). White quantization noise does, however, greatly ease subsequent analysis and comparison with the previously derived theoretical bounds.

One can improve the approximations by using higher order terms in various expansions of the quantizer nonlinearity, but this approach has not been noticeably successful in the ADC application, primarily because of its difficulty. In addition, traditional power series expansions are not well suited to the discontinuous nonlinearities of quantizers. (See Arnstein [2] and Slepian [75] for series expansion approximations for quantization noise in delta modulators and DPCM.) The approach taken here does, in fact, resemble this approach because the nonlinear operation will be expanded in a Fourier series, but all terms will be retained for an exact analysis. This general approach to nonlinearities in communication systems is a variation of the classical characteristic function method of Rice [70] and the transform method of Davenport and Root [25] who represented memoryless nonlinearities using Fourier or Laplace transforms. Here a Fourier series will play a similar role and the analysis will be for discrete time. A similar use of Fourier analysis for nonlinear systems was first applied to quantization noise analysis by Clavier, Panter, and Grieg in 1947 [20], [21]. Subsequently the characteristic function method was applied to the study of quantization noise by Widrow in 1960 [81] and Widrow's formulation has been used in subsequent development of conditions under which quantization noise is white, in particular by Sripad and Snyder [76] and Claasen and Jongepier [19].

6.2 Uniform Quantization

The basic common component of all ADCs is a quantizer. Uniform quantizers are popular because of their simplicity and because often there is no *a priori* knowledge of the statistical behavior of the quantizer input (or the original system input) and hence there is no way to optimize the quantizer levels for the input either using the high rate approximations of the previous chapter or using the iterative algorithms of the Lloyd-Max approach [61], [65], [30]. Furthermore, we have seen that the uniform quantizer is in fact approximately optimal in the high rate case when rate is measured by output entropy, that is, when subsequent noiseless coding is permitted.

Uniform quantizer are almost universally used in oversampled ADCs, systems that trade timing accuracy for quantizer complexity by replacing high rate quantization at the Nyquist rate by low rate (usually binary) quantization at many times the Nyquist rate. These latter systems are of interest here because they clearly violate the high rate assumption used for all of the high rate quantization bounds.

It is assumed that the quantizer has an even number, say M, of levels and that the distance between the output levels (the *bin width*) is Δ. A typical uniform quantizer is depicted in Figure 2.2. The quantizer output level is always the closest (nearest neighbor) level to the input so that

$$q(u) = \begin{cases} (\frac{M}{2} - \frac{1}{2})\Delta; \ (\frac{M}{2} - 1)\Delta \leq u \\ (k - \frac{1}{2})\Delta; \ (k-1)\Delta \leq u < k\Delta, \ k = (-\frac{M}{2} + 2), \ldots, (\frac{M}{2} - 1) \\ (-\frac{M}{2} + \frac{1}{2})\Delta; \ u < (-\frac{M}{2} + 1)\Delta \end{cases} \tag{6.2.1}$$

The quantizer error ϵ is defined by $\epsilon = q(u) - u$. Thus, if the input is a sequence of samples u_n, then we will be interested in the noise process $\epsilon_n = q(u_n) - u_n$ and we will wish to see if it resembles random white noise in some way.

Note that except for inputs in the top or bottom regions of the input range, the magnitude of the error ϵ is bound above by $\Delta/2$. If the input is confined to the range $[-M\Delta/2, M\Delta/2]$, then the magnitude error is guaranteed to be smaller than $\Delta/2$. If the input is outside of this range, then the quantizer magnitude error will be larger than $\Delta/2$ and we will say that the quantizer is *overloaded*. The range $[-M/2\Delta, M/2\Delta]$ is called the *granular region* or *no-overload range*. In an M level quantizer with bin width Δ, the no-overload range clearly has size $M\Delta$. Note that this use of the term "granular region" differs slightly from that of the general vector quantizer case, where only bounded quantizer cells were included. Here we also include a portion of the unbounded outermost cells and consider the granular region to be that region resulting in a quantizer error with magnitude less than $\Delta/2$ (which happens in a part of the unbounded regions).

In an ordinary scalar quantization or PCM system we can prevent quantizer overload by preceding the quantizer by a saturator. In a feedback quantization system, however, this makes even worse an already complicated analysis. It will be important, therefore, to determine conditions on a system under which the quantizer inside a feedback loop does not overload.

Normalize the quantizer output and input by the bin width Δ and use

6.2.1 to write

$$-\frac{\epsilon}{\Delta}=\frac{u}{\Delta}-\frac{q(u)}{\Delta} \tag{6.2.2}$$

$$=\begin{cases}\frac{u}{\Delta}-(\frac{M}{2}-1)-\frac{1}{2};\ (\frac{M}{2}-1)\le\frac{u}{\Delta}\\ \frac{u}{\Delta}-(k-1)-\frac{1}{2};\ (k-1)\le\frac{u}{\Delta}<k, k=(-\frac{M}{2}+2),\ldots,(\frac{M}{2}-1)\\ \frac{u}{\Delta}-(-\frac{M}{2}+1)+\frac{1}{2};\ \frac{u}{\Delta}<(-\frac{M}{2}+1)\end{cases}$$

If the input is confined to the no overload region, then the above formulas can be abbreviated to simply

$$-\frac{\epsilon}{\Delta}=\frac{u}{\Delta}-(k-1)-\frac{1}{2};\ (k-1)\le\frac{u}{\Delta}<k,\ k=(-\frac{M}{2}+1),\ldots,\frac{M}{2}. \tag{6.2.3}$$

This can be further abbreviated by introducing some notation. Every real number r can be uniquely written in the form $r=\lfloor r\rfloor+<r>$, where $\lfloor r\rfloor$ is the greatest integer less than or equal to r and $0\le<r><1$ is the fractional part of r (or r mod 1). From (6.2.3) we have that

$$-\frac{\epsilon}{\Delta}=\frac{u}{\Delta}-\lfloor\frac{u}{\Delta}\rfloor-\frac{1}{2}$$

and therefore

$$e=\frac{\epsilon}{\Delta}=\frac{1}{2}-<\frac{u}{\Delta}>. \tag{6.2.4}$$

The normalized quantizer error e is introduced here for convenience.

Figure 6.1 graphs e as a function of the quantizer u and makes it clear that for u constrained to the no-overload region, $e(u)$ is a periodic function of u with period 1.

Hence for u *in this region* we can write a Fourier series for e as

$$e=e(u)=\sum_{l\neq 0}\frac{1}{2\pi jl}e^{2\pi jl\frac{u}{\Delta}}=\sum_{l=1}^{\infty}\frac{1}{\pi l}\sin(2\pi l\frac{u}{\Delta}). \tag{6.2.5}$$

This series will hold for almost all values of u and it is the key formula for all of the subsequent analysis. Note that if we did not restrict u to lie in the no-overload region, then $e(u)$ would not be periodic and the Fourier series representation could not be used. Note also for later reference that one can similarly write a Fourier series for e^2 as

$$e(u)^2=\frac{1}{12}+\sum_{l\neq 0}\frac{1}{2(\pi l)^2}e^{2\pi jl\frac{u}{\Delta}}$$

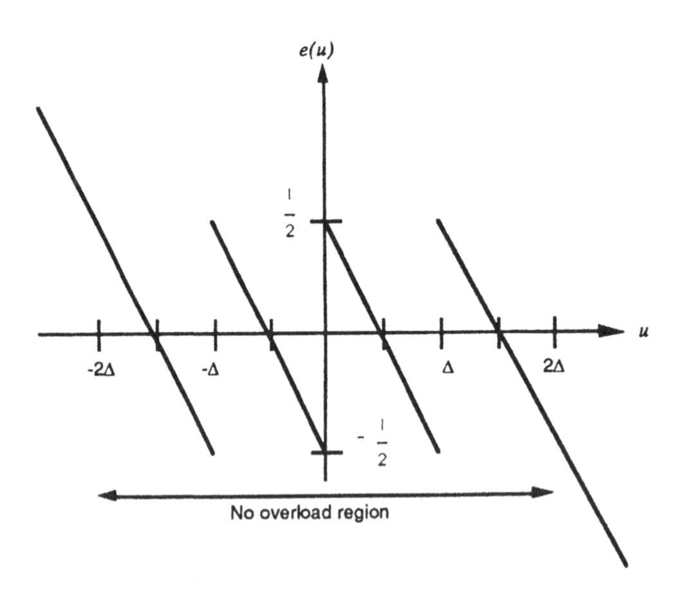

Figure 6.1: Quantizer Error

$$= \frac{1}{12} + \sum_{l=1}^{\infty} \frac{1}{2(\pi l)^2} \cos(2\pi l \frac{u}{\Delta}). \tag{6.2.6}$$

Suppose now that a sequence u_n is put into the quantizer, where we require that $|u_n| \leq M\Delta/2$ so that there is no overload. We wish to study the behavior of the normalized error sequence $e_n = \epsilon_n/\Delta$. Here we will follow two different, but related, approaches. The first is that of Clavier, et al. [20], [21]: From (6.2.5) it is immediate that

$$e_n = \frac{1}{2} - < u_n >= \sum_{l \neq 0} \frac{1}{2\pi j l} e^{2\pi j l \frac{u_n}{\Delta}} = \sum_{l=1}^{\infty} \frac{1}{\pi l} \sin(2\pi l \frac{u_n}{\Delta}). \tag{6.2.7}$$

For some specific examples of sequences u_n, the above formula can be used to obtain a form of Fourier series representation directly for the sequence e_n. This representation then can be used to determine moments and spectral behavior. One problem with this approach is that it only works for certain simple inputs, e.g., sinusoids. Another difficulty is that ordinary Fourier series may not work (i.e., converge) since e_n need not be a periodic function of n. For example, if $u_n = A\cos(2\pi f_0 n)$ and f_0 is not a rational number, then u_n (and hence also e_n) is not periodic. (For example, $u_n = 0$ only if $n = 0$.) This problem can be handled mathematically by resorting to a generalization of the Fourier series due to Harald Bohr and the related theory of almost periodic functions, but we shall only make a few relevant observations when appropriate. (For a complete treatment of almost periodic functions, Bohr's classic [10] still makes good reading. See also [8], [9] and the discussion in [68] and the references therein.)

An alternative approach which is also based on (6.2.5)–(6.2.6) is to instead focus on moments of the process rather than on a specific Fourier representation of the actual sequence. This will lead to the characteristic function method [70]. Here the primary interest is the long term statistical behavior of the error sequence e_n. We assume that the sequences are quasi-stationary as defined in Chapter 1 and consider the mean, autocorrelation function, and second moment defined by

$$m_e = \bar{E}\{e_n\} \tag{6.2.8}$$

$$R_e(k) = \bar{E}\{e_n e_{n+k}\}, \tag{6.2.9}$$

$$R_e(0) = \bar{E}\{e_n^2\}. \tag{6.2.10}$$

These moments reduce to the corresponding time averages or probabilistic averages in the special cases of deterministic or random processes, respectively.

We now proceed to apply the basic formulas (6.2.7) and (6.2.6) to find an expression for the basic moments of (6.2.9)–(6.2.10). Plugging (6.2.7) into (6.2.8) and (6.2.6) into (6.2.10) and assuming that the limits can be interchanged results in

$$\bar{E}\{e_n\} = \sum_{l\neq 0} \frac{1}{2\pi j l} \bar{E}\{e^{2\pi j l \frac{u_n}{\Delta}}\},$$

$$\bar{E}\{e_n^2\} = \frac{1}{12} + \sum_{l\neq 0} \frac{1}{2(\pi l)^2} \bar{E}\{e^{2\pi j l \frac{u_n}{\Delta}}\},$$

and for $k \neq 0$

$$R_e(k) = \sum_{i\neq 0} \sum_{l\neq 0} \frac{j}{2\pi i} \frac{j}{2\pi l} \bar{E}\{e^{2\pi j (i \frac{u_n}{\Delta} + l \frac{u_{n+k}}{\Delta})}\}$$

These expressions can be most easily given in terms of the one-dimensional characteristic function

$$\bar{\Phi}_u(l) = \bar{E}\{e^{2\pi j l \frac{u_n}{\Delta}}\} \tag{6.2.11}$$

and a two-dimensional characteristic function

$$\bar{\Phi}_u^{(k)}(i, l) = \bar{E}\{e^{2\pi j (i \frac{u_n}{\Delta} + l \frac{u_{n+k}}{\Delta})}\}, k \neq 0, \tag{6.2.12}$$

as

$$\bar{E}\{e_n\} = \sum_{l\neq 0} \frac{1}{2\pi j l}\bar{\Phi}_u(l), \tag{6.2.13}$$

$$\bar{E}\{e_n^2\} = \frac{1}{12} + \sum_{l\neq 0} \frac{1}{2(\pi l)^2}\bar{\Phi}_u(l), \tag{6.2.14}$$

and for $k \neq 0$

$$R_e(k) = -\sum_{i\neq 0}\sum_{l\neq 0} \frac{1}{2\pi i}\frac{1}{2\pi l}\bar{\Phi}_u^{(k)}(i,l). \tag{6.2.15}$$

The interchange of the limits is an important technical point that must be justified in any particular application. For the purely deterministic case, it can be shown that if the one dimensional time-average characteristic function of (6.2.11) exists (i.e., the limit converges to something finite and $PHI_u(l) \to 0$ as $l \to \infty$), then so does the two-dimensional and the above limit interchanges are valid [39]. If the process is stationary, then the characteristic functions reduce to the usual probabilistic characteristic functions and a similar conclusion follows.

If the characteristic functions of (6.2.11)–(6.2.12) can be evaluated, then the moments and spectrum of the process can be computed from (6.2.13)–(6.2.15). This general approach to finding the moments of the output of a nonlinear system is a variation of the "characteristic function" approach described by Rice [70] and studied by Davenport and Root [25] under the name "transform method." Here discrete time replaces continuous time, Fourier series replace Laplace transforms, and stationary processes are replaced by quasi-stationary processes (and hence time-averages can replace probabilistic averages).

6.3 PCM Quantization Noise: Deterministic Inputs

We first consider a purely deterministic input and hence the averages will all be time averages, that is, $\bar{E}\{x_n\} = M\{x_n\}$. We will not consider in detail the example of a dc input to an ordinary uniform quantizer in any detail because the results are trivial: If the input is dc, say $u_n = u$, then the error is simply $e_n = \epsilon_n/\Delta = \epsilon(u)/\Delta = 1/2 - u/\Delta$ and the mean is this constant value and the second moment is the constant squared (and hence the variance is 0). The autocorrelation is also equal to the second moment. This does point out, however, that the common white noise assumption is certainly erroneous for a dc input since $R_e(k)$ is not 0 for $k \neq 0$. This is well known and it is generally believed that the input to the quantizer must be "active" if the noise is to be approximately white. Hence we consider next a more interesting input, a sinusoid $u_n = A\sin(n\omega_0 + \theta)$ with a fixed initial phase θ.

The dc case for simple PCM does become interesting when dithering is used. This case is treated in the exercises.

We assume that $A \leq M/2$ so that the quantizer is not overloaded. Define also $f_0 = \omega_0/2\pi$. Here we can directly use (6.2.7) as did Clavier, et al. [20], [21] to find an expression for the error sequence. For the given case,

$$e_n = \sum_{l \neq 0} \frac{1}{2\pi j l} e^{2\pi j l \frac{A}{\Delta} \sin(n\omega_0 + \theta)}. \tag{6.3.1}$$

For brevity we denote A/Δ by γ. To evaluate this sum we use the Jacobi-Anger formula

$$e^{jz\sin\psi} = \sum_{m=-\infty}^{\infty} J_m(z) e^{jm\psi}, \tag{6.3.2}$$

where J_m is the ordinary Bessel function of order m to obtain

$$e_n = \sum_{m=-\infty}^{\infty} e^{jn\omega_0 m} \left(e^{jm\theta} \sum_{l \neq 0} \frac{1}{2\pi j l} J_m(2\pi l \gamma) \right)$$

$$= \sum_{m=-\infty}^{\infty} e^{jn\omega_0(2m-1)} \left(e^{j(2m-1)\theta} \sum_{l=1}^{\infty} \frac{1}{\pi j l} J_{2m-1}(2\pi l \gamma) \right), \tag{6.3.3}$$

where the even terms disappear because $J_m(z) = (-1)^m J_m(-z)$. Eq. (6.3.3) provides a generalized Fourier series representation for the sequence e_n in the form

$$e_n = \sum_{m=-\infty}^{\infty} e^{j\lambda_m n} b_m, \tag{6.3.4}$$

where the qualifier "generalized" is used because the sequence e_n need not be periodic. From this expression one can identify the power spectrum of e_n as having components of magnitude $|b_m|^2$ at frequencies $\lambda_m = (2m-1)\omega_0 \bmod 2\pi$. Note that if f_0 is rational, then the input and hence also e_n will be periodic and the frequencies $(2m-1)\omega_0 \bmod 2\pi$) will duplicate as the index m takes on all integer values. In this case, the power spectrum amplitude at a given frequency is found by summing up all of the coefficients $|b_m|^2$ with $(2m-1)\omega_0 \bmod 2\pi$ giving the desired frequency. We will henceforth assume that f_0 is not a rational number (and hence that in the corresponding continuous time system, the sinusoidal frequency is incommensurate with the sampling frequency). This assumption simplifies the analysis and can be physically justified since a randomly selected frequency from a continuous probability distribution will be irrational with probability one. An irrational ω_0 can also be viewed as an approximation to a rational frequency with a large denominator when expressed in lowest terms.

The above direct approach indeed gives the spectrum for this special case. In particular, from the symmetry properties of the Bessel functions one has a power spectrum that assigns amplitude

$$S_m = \left(\frac{1}{\pi}\sum_{l=1}^{\infty}\frac{1}{l}J_{2m-1}(2\pi\gamma l)\right)^2 \tag{6.3.5}$$

to every frequency of the form $(2m-1)\omega_0 \bmod 2\pi)$. Letting $\delta(f)$ denote a Dirac delta function, this implies a power spectral density of the form

$$S_e(f) = \sum_{m=-\infty}^{\infty} S_m \delta(f - < (2m-1)f_0 >). \tag{6.3.6}$$

We could use these results to evaluate the mean, power, and autocorrelation. Instead we turn to the more general transform method. The two methods will be seen to yield the same spectrum for this example.

For the given purely deterministic example, the one-dimensional characteristic function can be expressed as

$$\bar{\Phi}_u(l) = \lim_{N\to\infty}\frac{1}{N}\sum_{n=1}^{N} e^{j2\pi l\gamma\sin(n\omega_0+\theta)}$$

$$= \lim_{N\to\infty}\frac{1}{N}\sum_{n=1}^{N} e^{j2\pi l\gamma\sin(2\pi<nf_0+\frac{\theta}{2\pi}>)}, \tag{6.3.7}$$

where the fractional part can be inserted since $\sin(2\pi u)$ is a periodic function in u with period 1. This limit can be evaluated using a classical result in ergodic theory of Hermann Weyl (see, e.g., Petersen [68]): If g is an integrable function, a is an irrational number, and b is any real number, then

$$\lim_{N\to\infty}\frac{1}{N}\sum_{n=1}^{N} g(< an+b >) = \int_0^1 g(u)du. \tag{6.3.8}$$

This remarkable result follows since the sequence of numbers $< an+b >$ uniformly fills the unit interval and hence the sums approach an integral in the limit. Applying (6.3.8) to (6.3.7) yields

$$\bar{\Phi}_u(l) = \int_0^1 du e^{j2\pi l\gamma \sin(2\pi u)} = J_0(2\pi l\gamma), \tag{6.3.9}$$

The mean and second moment of the quantizer noise can then be found using the fact $J_0(r) = J_0(-r)$:

$$\bar{E}\{e_n\} = \sum_{l\neq 0} \frac{1}{2\pi jl} J_0(2\pi l\gamma) = 0, \tag{6.3.10}$$

$$\bar{E}\{e_n^2\} = \frac{1}{12} + \frac{1}{\pi^2}\sum_{l=1}^{\infty} \frac{1}{l^2} J_0(2\pi l\gamma). \tag{6.3.11}$$

Note that the result does not depend on the frequency of the input sinusoid and that the time average mean is 0, which agrees with that predicted by the assumption that ϵ_n is uniformly distributed on $[-\Delta/2, \Delta/2]$. The second moment, however, differs from the value of 1/12 predicted by the uniform assumption by the right hand sum of weighted Bessel functions. No simple closed form expression for this sum is known to the author, but other formulas useful for insight and computation can be developed. Before pursuing these, note that if $\gamma = A/\Delta$ becomes large (which with A held fixed and the no-overload assumption means that the number of quantization levels is becoming large and the bin-width Δ is becoming small), then $J_0(2\pi l\gamma) \to 0$ and hence the second moment converges to 1/12 in the limit. This is consistent with high rate quantization theory.

By expressing the Bessel function as an integral of a cosine and by applying the summation

$$\sum_{k=1}^{\infty} \frac{\cos kx}{k^2} = \sum_{k=1}^{\infty} \frac{\cos 2\pi k < \frac{x}{2\pi} >}{k^2} = \pi^2 (\frac{1}{6} - < \frac{x}{2\pi} > + < \frac{x}{2\pi} >^2) \tag{6.3.12}$$

(the second equality following, e.g., from Gradshteyn and Rhyzhik [34]) to obtain

$$\bar{E}\{e_n^2\} = \frac{1}{\pi}\int_0^{\pi} (\frac{1}{2} - < \gamma \sin\psi >)^2 \, d\psi$$

$$= \frac{2}{\pi}\int_0^{\frac{\pi}{2}} (\frac{1}{2} - < \gamma \sin\psi >)^2 \, d\psi. \tag{6.3.13}$$

This result can also be obtained by applying Weyl's result directly to $\bar{E}\{e_n^2\}$ without using the transform approach. An alternative form can be found by observing that since $<\gamma\sin\psi>=\gamma\sin\psi-(k-1)$ when $k>\gamma\sin\psi\geq k-1$, we have that if $K=\lceil\gamma\rceil$, the smallest interger greater than or equal to K,

$$\bar{E}\{e_n^2\}=\frac{2}{\pi}\sum_{k=1}^{K}\int_{b_{k-1}}^{b_k}(k-\frac{1}{2}-\gamma\sin\psi)^2d\psi,$$

where $b_0=0$, $b_k=\sin^{-1}(k\Delta/A)$, $k=1,2,\ldots,K-1$, $b_K=\pi/2$. With some algebra this yields

$$\bar{E}\{e_n^2\}=(K+\frac{1}{2})^2+\gamma^2+\frac{2}{\pi}\frac{A}{\Delta}-2-\frac{2}{\pi}(\sum_{k=1}^{K-1}\sin^{-1}(\frac{k\Delta}{A})+\sum_{k=1}^{K-1}\cos(\sin^{-1}(\frac{k\Delta}{A}))). \tag{6.3.14}$$

The right-most term can be further simplified using the fact that

$$\cos(\sin^{-1}(a))=\sqrt{1-a^2}.$$

Eq. (6.3.14) has the advantage over (6.3.11) that it is expressed as a finite sum and it more resembles the form of the results of [20] for the harmonic distortion when a uniform quantizer with an odd number of levels is driven by a full range sinusoidal input.

To compute the autocorrelation of the quantization noise, we use similar steps to find the joint characteristic function $\bar{\Phi}_u^{(k)}(i,l)$. We have that

$$\bar{\Phi}_u^{(k)}(i,l)=\lim_{N\to\infty}\frac{1}{N}\sum_{n=1}^{N}e^{2\pi j[l\gamma\sin(n\omega_0+\theta)+i\gamma\sin((n+k)\omega_0+\theta)]}$$

$$=\lim_{N\to\infty}\frac{1}{N}\sum_{n=1}^{N}e^{2\pi j[l\gamma\sin(2\pi<nf_0+\frac{\theta}{2\pi}>)+i\gamma\sin(2\pi<nf_0+\frac{\theta}{2\pi}>+k\omega_0)]}$$

$$=\int_0^1 due^{2\pi j[l\gamma\sin(2\pi u)+i\gamma\sin(2\pi u+k\omega_0)]}. \tag{6.3.15}$$

Using the Jacobi-Anger formula and the orthogonality of exponentials yields

$$\bar{\Phi}_u^{(k)}(i,l)=\sum_{n=-\infty}^{\infty}J_n(2\pi\gamma i)J_{-n}(2\pi\gamma l)e^{jnk\omega_0}. \tag{6.3.16}$$

Inserting (6.3.15) into (6.2.15) and using the fact that $J_{-n}(x)=(-1)^nJ_n(x)$ yields

$$R_e(k)=-\frac{1}{(2\pi)^2}\sum_{n=-\infty}^{\infty}(-1)^n(\sum_{l\neq 0}\frac{J_n(2\pi\gamma l)}{l})^2e^{jnk\omega_0}$$

$$= \sum_{n=-\infty}^{\infty} (\frac{1}{\pi} \sum_{l=1}^{\infty} \frac{J_{2n-1}(2\pi\gamma l)}{l})^2 e^{jnk\omega_0}. \tag{6.3.17}$$

This has the form

$$R_e(k) = \sum_{n=-\infty}^{\infty} S_n e^{2\pi jk\lambda_n}, \tag{6.3.18}$$

where $\lambda_n =< (2n-1)\omega_0/2\pi >$ are normalized frequencies in $[0,1)$ and

$$S_n = (\frac{1}{\pi} \sum_{l=1}^{\infty} \frac{J_{2n-1}(2\pi\gamma l)}{l})^2 \tag{6.3.19}$$

are the spectral components at the frequency λ_n, and hence

$$S_e(f) = \sum_n S_n \delta(f - \lambda_n), \tag{6.3.20}$$

exactly as found directly in (6.3.5)–(6.3.6).

The sequence S_m is sometimes referred to as the *Bohr-Fourier* series of the autocorrelation because of Bohr's generalization of the ordinary Fourier series of periodic functions to the general exponential decomposition of (6.3.18), which implies that $R_e(k)$ is an almost periodic function of k. It suffices here to point out that the spectrum of the quantizer noise is purely discrete and consists of all odd harmonics of the fundamental frequency of the input sinusoid. The energy at each harmonic depends in a very complicated way on the amplitude of the input sinusoid. In particular, the quantizer noise is decidedly not white since it has a discrete spectrum and since the spectral energies are not flat. Thus here the white noise approximation of Bennett and of the describing function approach is invalid, even if M is large.

An alternative form for (6.3.19) involving only a finite sum can be found in the same way (6.3.14) was found from (6.3.11). Since for odd n

$$J_n(z) = \frac{2}{\pi} \int_0^{\frac{\pi}{2}} \sin(n\theta) \sin(z \sin\theta)\, d\theta \tag{6.3.21}$$

we have, using the summation [34]

$$\sum_{l=1}^{\infty} \frac{\sin(xl)}{l} = \frac{\pi}{2} - \frac{1}{2}x;\ 0 \le x < 2\pi, \tag{6.3.22}$$

that for odd n

$$\sum_{l=1}^{\infty} \frac{J_n(2\pi\gamma l)}{l} = \frac{2}{\pi} \int_0^{\frac{\pi}{2}} d\theta \sin(n\theta) \sum_{l=1}^{\infty} \frac{\sin(2\pi < \gamma \sin\theta > l)}{l}$$

$$= \frac{2}{\pi}\int_0^{\frac{\pi}{2}} d\theta \sin(n\theta)(\frac{\pi}{2} - \pi < \gamma\sin(\theta) >)$$

and hence as in (6.3.14)

$$\sum_{l=1}^{\infty}\frac{J_n(2\pi\gamma l)}{l} = \frac{1}{n} - 2\gamma\frac{\pi}{4}\delta_{n-1} + 2\sum_{k=1}^{K}\int_{b_{k-1}}^{b_k} d\theta\sin(n\theta)(\gamma\sin\theta - (k-1)) \tag{6.3.23}$$

$$= \frac{1}{n} - 2\gamma\frac{\pi}{4}\delta_{n-1} + 2\sum_{k=1}^{K} k\int_{b_{k-1}}^{b_k} \sin(n\theta)\,d\theta = -\frac{2}{\pi n}\sum_{k=0}^{K-1}\cos(n\sin^{-1}(\frac{\Delta}{A}k)),$$

where, as previously, $K \le M/2$ is the largest integer greater than or equal to γ. Thus (6.3.19) can also be written as

$$S_n = (\frac{1}{n} - 2\gamma\frac{\pi}{4}\delta_{n-1} + \frac{2}{\pi n}\sum_{k=0}^{K-1}\cos((2n-1)\sin^{-1}(\frac{\Delta}{A}k)))^2. \tag{6.3.24}$$

Eq. (6.3.24) strongly resembles the corresponding result of Clavier, et al., for the case of odd M and a full range sinusoidal input (that is, $\gamma = A/\Delta = (M-1)/2$).

6.4 Random Inputs and Dithering

As a simple observation that will not be worked out in detail, an obvious variation of the deterministic example of the previous section is to consider a process $u_n = A\sin(2\pi f_0 + \theta)$ with θ a uniform random variable. This transforms the deterministic example into a stationary random process and the moments become probabilistic averages instead of time averages. The final answers are the same, although the input frequency need not be assumed to be irrational.

Another simple example is the case of an iid input uniformly distributed over the no-overload range. In this case it is well known (and easy to prove) that the quantization noise is exactly an iid process uniformly distributed over $[-\Delta/2, \Delta/2]$.

A much more interesting example is a process of the form

$$u_n = v_n + w_n, \tag{6.4.1}$$

where v_n is the possibly nonstationary original system input (such as the deterministic sinusoid previously considered) and w_n is an iid random process which is called a *dither* signal. A key attribute of the dither signal is that it is independent of the v_n process, that is, v_n is independent of

w_k for all times n and k. We still require that the quantizer input u_n be in the no-overload region. This has the effect of reducing the allowed input dynamic range and hence limiting the overall SQNR. Dithering has long been used as a means of improving the subjective quality of quantized speech and images (see Jayant and Noll [52], Section 4.8, and the references therein). The principal theoretical property of dithering was developed by Schuchman [71] who proved that if the quantizer does not overload and the characteristic function of the marginal probability density function of the dither signal is 0 at integral multiples of $2\pi/\Delta$, then the quantizer error e_n (at a specific time n) is independent of the original input signal x_k for all k. (We will prove this result shortly.) It is generally thought that a good dither signal should force the quantizer error to be white even when the input signal has memory (the quantizer noise is trivially white if the original input is white). General results on this popular belief follow from the work of Sripad and Snyder [76], but this specific result is not explicitly stated by them. It is easily proved using the characteristic function method.

Given an input as in (6.4.1) with w_n a stationary random process, let

$$\Phi_w(j\alpha) = E(e^{j\alpha w_n})$$

denote the ordinary characteristic function of w_n (which does not depend on n since $\{w_n\}$ is stationary). Because of the independence of the processes, the one-dimensional characteristic function of (6.2.11) becomes

$$\bar{\Phi}_u(l) = \lim_{N\to\infty} \frac{1}{N} \sum_{n=1}^{N} E(e^{2\pi j l \frac{1}{\Delta}(v_n + w_n)}) = \Phi_w(j2\pi\frac{l}{\Delta})\bar{\Phi}_v(l). \tag{6.4.2}$$

The two-dimensional characteristic function of (2.18) is

$$\begin{aligned}\bar{\Phi}_u^{(k)}(i,l) &= \lim_{N\to\infty} \frac{1}{N} \sum_{n=1}^{N} E(e^{2\pi j \frac{1}{\Delta}[i(v_n+w_n)+l(v_{n+k}+w_{n+k})]}) \\ &= \Phi_w(2\pi j\frac{1}{\Delta}i)\Phi_w(2\pi j\frac{1}{\Delta}l)\bar{\Phi}_v^{(k)}(i,l);\, k \neq 0. \end{aligned} \tag{6.4.3}$$

Now suppose that the marginal distribution of w_n is such that Schuchman's conditions [71] are satisfied, that is, the quantizer is not overloaded and

$$\Phi_w(2\pi j\frac{1}{\Delta}l) = 0; l = 1, 2, 3, \ldots. \tag{6.4.4}$$

(Recall that $\Phi_w(0) = 1$ for any distribution.) This is the condition shown by Schuchman to be necessary and sufficient for the quantization error to be independent of the original input x_n. The principal example is a dither signal with a uniform marginal on $[-\Delta/2, \Delta/2]$ (and an input amplitude constrained to avoid overload when added to the dither), in which case

$$\Phi_w(j\alpha) = \frac{2}{\alpha\Delta} \sin \frac{\alpha\Delta}{2}. \tag{6.4.5}$$

For this case we have

$$\bar{\Phi}_u(l) = \begin{cases} 1; & l = 0 \\ 0; & \text{otherwise.} \end{cases} \tag{6.4.6}$$

and for $k \neq 0$

$$\bar{\Phi}_u^{(k)}(i, l) = \begin{cases} \bar{\Phi}_v^{(k)}(0, 0) = 1; & i = l = 0 \\ 0; & \text{otherwise.} \end{cases} \tag{6.4.7}$$

Thus in this example we have from (6.2.13)–(6.2.15) that e_n has zero mean, a second moment of $1/12$, and an autocorrelation function $R_e(k) = 0$ when $k \neq 0$, that is, the quantization noise is indeed white as is generally believed when Schuchman's condition is satisfied. Note that this is true for a general quasi-stationary input, including the sinusoid previously considered. Thus, for example, if the input sequence is $A\sin(n\omega_0)$ as before and the dither sequence is an iid sequence of uniform random variables on $[-\Delta/2, \Delta/2]$, then the no-overload condition becomes $A + \Delta/2 \leq \Delta M/2$ or

$$\frac{A}{\Delta} \leq \frac{M-1}{2}, \tag{6.4.8}$$

which has effectively reduced the allowable γ. Thus, for example, if $M = 8$ as in Fig. 2.2, the maximum allowable γ is reduced from 4 to 7/2.

A similar analysis can be used to prove Schuchman's original result that the error and signal are independent. We here focus on a weaker result that is more in tune with the development thus far and show that the two processes are uncorrelated. Consider the cross-correlation

$$R_{ev}(k) = \bar{E}\{e_n v_{n+k}\} = \bar{E}\{v_{n+k} \sum_{l \neq 0} e^{2\pi jl \frac{v_n + w_n}{\Delta}}\}.$$

Since the dither signal w_n is assumed to be independent of the input v_n, this becomes

$$R_{ev}(k) = \sum_{l \neq 0} \bar{E}\{v_{n+k} e^{2\pi jl \frac{v_n}{\Delta}}\} \Phi_w(2\pi j \frac{l}{\Delta}),$$

where we have assumed that the sum and limit can be interchanged. Thus if Schuchman's conditions 6.4.4 are satisfied,

$$R_{ev}(k) = 0; \quad \text{all } k,$$

as claimed.

Although the addition of a dither signal does turn the quantization noise into a simpler process and decorrelates the noise from the signal, it should be remembered that this is not the goal of a quantizer. Adding a dither signal also adds distortion to the final reproduction since it corrupts the original signal in a possibly non-invertible manner. It also reduces the SQNR achievable with a given quantizer since the input amplitude must be reduced enough so that the original signal plus the dither stays within the no-overload region. This loss may be acceptable (and small) when the number of quantization levels is large. It is significant if there are only a few quantization levels. For example, if $M = 2$, then a uniform dither on $[-\Delta/2, \Delta/2]$ can only avoid overload if the signal is a 0 dc signal.

The statement that dithering corrupts a source can be tempered if one assumes that the dithering signal is not really a random signal, but is instead a pseudo-random signal known both to the encoder and decoder. One still assumes it is random in the previous analysis, but one now also assumes that the decoder has a perfect copy of the dithering signal. In this case the previous analysis implies that

$$q(u_n + w_n) = u_n + w_n + e_n$$

with e_n independent of u_n (since the quantization noise is independent of the signal from Schuchman's result). Thus if we now remove w_n from $q(u_n + w_n)$ we have that

$$q(u_n + w_n) - w_n = u_n + e_n,$$

that is, the decoded signal is just the original signal plus noise that we proved to be uniform and white. Thus in the case of "subtracted" dither, you can have your cake and eat it too: the dither whitens the noise and makes it uniform (thereby simplifying noise analysis), but it does not corrupt the signal! There remains a penalty, however, in that the input range has been reduced in order to avoid overloading the quantizer.

As a final remark, one could reverse the roles of the two processes and consider an original input to be a stationary process w_n and v_n to be a sinusoidal dither, as proposed by Jaffee [51]. Here the two dimensional probabilistic characteristic function of the stationary process factors out of the time average of the sinusoidal dither. In general, however, the resulting characteristic functions will not be 0 for nonzero arguments and the noise will not be white.

6.5 Sigma-Delta Modulation

The basic Sigma-Delta modulator can be motivated by an intuitive argument based on the dithering idea. Suppose that instead of adding an iid random process to the signal before quantization, the quantization noise itself is used as a dither signal, that is, iid signal-independent noise is replaced by deterministic signal-dependent noise which (hopefully) approximates a white signal-independent process. The idea of using a deterministic dither signal drawn from the quantization noise in PCM is generally attributed to C. C. Cutler, who first patented differential PCM in 1952. Reversing the noise sign for convenience and inserting a delay in the forward path of the feedback loop (to reflect the physical delay inherent in a quantizer) yields the system of Figure 6.2, where we preserve the u_n notation for the quantizer input and label the original system input as x_n.

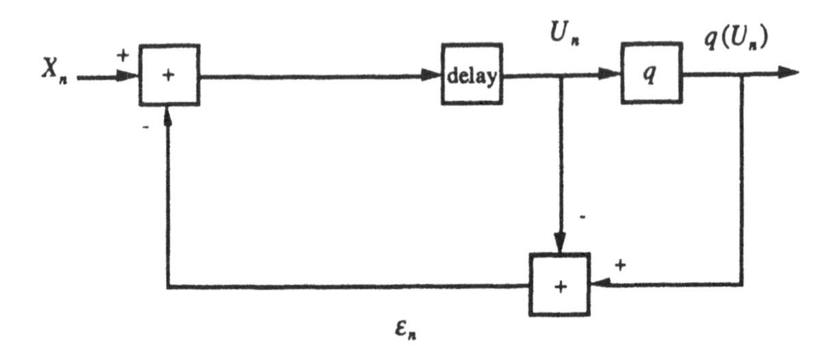

Figure 6.2: Deterministic Dithering

The nonlinear difference equation governing this system is

$$u_n = x_{n-1} - \epsilon_{n-1} = u_{n-1} + x_{n-1} - q(u_{n-1});\ n = 1, 2, \ldots \qquad (6.5.1)$$

This difference equation can also be depicted as in Figure 6.3, which is the traditional configuration for a single-loop Sigma-Delta modulator.

The name follows from the fact that the system can be considered as a delta modulator preceeded by an integrator. The system was originally called a Delta-Sigma modulator [49] and this terminology is also common. The modern popularity of these systems, much of the original analysis, and the name Sigma-Delta modulator is due to Candy and his colleagues [12], [13], [14], [16], [15].

One might hope that the deterministic dither might indeed yield a white quantization noise process, but unfortunately this circular argument does

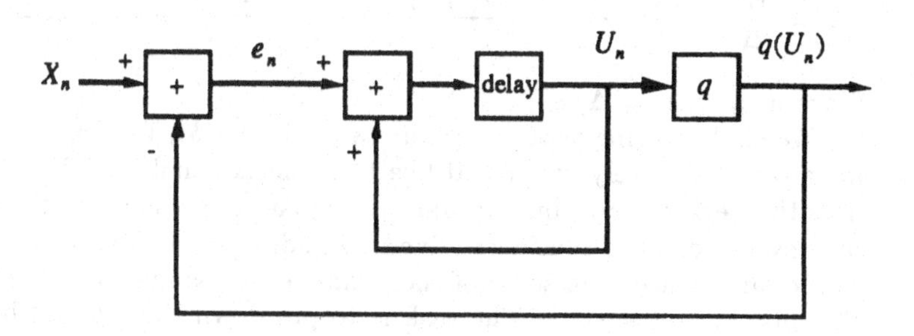

Figure 6.3: Traditional Sigma-Delta Modulator

not hold for the simple system of Figure 6.2 or 6.3. In this section we develop the moments and spectra for the basic Sigma-Delta modulator and evaluate the results for the special case of a dc input. This case is of interest in this example as Sigma-Delta modulation is usually only used in oversampled ADCs, that is, the input signal is sampled at many times the Nyquist rate and hence the input is very slowly varying, appearing like a dc over many samples. The analysis can be extended to sinusoidal inputs.

It should be pointed out that here we will be concerned only with the behavior of the quantizer noise, that is, the difference between the input and output of the quantizer inside the feedback loop. We do not tackle the problem of evaluating the overall distortion between the original input and final output of the ADC. The emphasis here is on the behavior of the quantizer itself and on comparing this behavior with the asymptotic approximations of the previous chapter. Knowing the quantizer noise behavior one can find or bound the overall distortion. The interested reader is referred to the cited literature for details.

Since $u_n = q(u_n) - \epsilon_n$, (6.5.1) yields the difference equation

$$q(u_n) = x_{n-1} + \epsilon_n - \epsilon_{n-1}, \tag{6.5.2}$$

which has the intuitive interpretation that the quantizer output can be written as the input signal (delayed) plus a difference (or discrete time derivative) of an error signal. The hope is that this difference will be a high frequency term which can be removed by low pass filtering to obtain the original signal. For convenience we assume that $u_0 = 0$ and we normalize the above terms by Δ and use the definition of ϵ_n to write

$$e_n = \frac{\epsilon_n}{\Delta} = \frac{q(u_n) - u_n}{\Delta} = \frac{q(x_{n-1} - \epsilon_{n-1})}{\Delta} - \frac{x_{n-1}}{\Delta} - e_{n-1};\ n = 1, 2, \ldots. \tag{6.5.3}$$

Since $u_0 = 0$, $\epsilon_0 = \Delta/2$.

We shall assume that the input range is $[-b, b)$, that is, $-b \leq x_n < b$ for all n. Intuitively, we would like to make Δ small in order to keep the quantizer error small; but we dare not make it too small or the quantizer may overload (M is considered fixed). In addition, we do not want to make Δ too large since then some of the quantizer levels may never be used and the bits will be wasted. The task is to pin down the relation between Δ and b, that is, given b, how do we choose Δ? This is the key property of the system and is found using an induction argument. We wish to choose Δ so that the quantizer will never overload as long as the input stays within its range. A necessary condition is immediate: If $\epsilon_0 = \Delta/2$, then to ensure that $|u_1| \leq M\Delta/2$ we must have that

$$|x_0 - \frac{\Delta}{2}| \leq \frac{M\Delta}{2}$$

and therefore that

$$b + \frac{\Delta}{2} \leq \frac{M\Delta}{2}$$

or

$$b \leq (M-1)\frac{\Delta}{2}. \tag{6.5.4}$$

We now show that this condition is sufficient to ensure that the quantizer does not overload; that is, if Δ satisfies (6.5.4), then $|\epsilon_n| \leq \Delta/2$ for all n. We have already seen this is true for $n = 1$, so we proceed by induction. Assume that it is true for $k = 1, 2, \ldots, n-1$. If $u_n = x_{n-1} - \epsilon_{n-1}$ is in the no-overload region $[-M\Delta/2, M\Delta/2]$ then $|\epsilon_n| \leq \Delta/2$ by definition of no overload. Thus we need only prove that it is not possible for either $u_n > M\Delta/2$ or $u_n < -M\Delta/2$. We have, however, from the range of x_{n-1} and the assumption on ϵ_{n-1} that

$$-b - \frac{\Delta}{2} \leq x_{n-1} - \epsilon_{n-1} \leq b + \frac{\Delta}{2}$$

and hence using (6.5.4) we have that

$$|u_n| = |x_{n-1} - \epsilon_{n-1}| \leq b + \frac{\Delta}{2} \leq \frac{M\Delta}{2}, \tag{6.5.5}$$

proving that the quantizer does not overload. Observe that the input bin for the largest quantizer level is $\{u : u \geq (M/2 - 1)\Delta\}$. Thus this level will be selected for $u_1 = x_0 - \Delta/2$ if $x_0 - \Delta/2 \geq (M/2-1)\Delta$ or $x_0 \geq (M/2-1/2)\Delta$. If $b < (M-1)\Delta/2$ this will never happen and the level will not be used. Combining this observation with (6.5.4) we are led to the following choice for Δ:

$$\Delta = \frac{2b}{M-1}. \tag{6.5.6}$$

In the most common case of a binary quantizer, $\Delta = 2b$. We can now combine the fact that the quantizer does not overload with the formula (6.2.4) for the quantizer error for a nonoverloading quantizer to obtain a recursion for the quantizer error sequence of a single-loop Sigma-Delta quantizer: $e_0 = 1/2$,

$$e_n = \frac{1}{2} - < \frac{u_n}{\Delta} > = \frac{1}{2} - < \frac{x_{n-1}}{\Delta} - e_{n-1} >; \; n = 1, 2, \ldots \tag{6.5.7}$$

This formula can now be solved by induction so that the e_n terms on the right are eliminated. It is simpler to express the recursion in terms of

$$y_n = \frac{1}{2} - e_n = < \frac{x_{n-1}}{\Delta} + \frac{1}{2} + y_{n-1} >,$$

the solution to which is

$$y_0 = 0$$

$$y_n = < \sum_{k=0}^{n-1} \left(\frac{1}{2} + \frac{x_k}{\Delta}\right) > = < \frac{n}{2} + \sum_{k=0}^{n-1} \frac{x_k}{\Delta} >, n = 1, 2, \ldots \tag{6.5.8}$$

as can be verified by induction. Thus the quantizer error sequence is given by

$$e_n = \frac{1}{2} - < \frac{n}{2} + \sum_{k=0}^{n-1} \frac{x_k}{\Delta} >, \tag{6.5.9}$$

which provides an interesting comparison with the corresponding expression for the quantizer error sequence of an ordinary uniform quantizer operating on an input sequence u_n of (6.2.4), that is,

$$e_n = \frac{1}{2} - < \frac{u_n}{\Delta} > .$$

Thus when the quantizer is put into a feedback loop with an integrator, the overall effect is to integrate the input plus a constant bias before taking the fractional part. The overall nonlinear feedback loop therefore appears as an affine operation (linear plus a bias) on the input followed by a memoryless nonlinearity. Furthermore, the techniques used to find the time average moments for e_n in the memoryless quantizer case can now be used by replacing u_n by the sum

$$s_n = \sum_{k=0}^{n-1} (\frac{1}{2} + \frac{x_k}{\Delta}), \tag{6.5.10}$$

evaluating the characteristic functions of (6.2.11)–(6.2.12) and applying (6.2.13)–(6.2.15) for $\bar{\Phi}_s$ instead of $\bar{\Phi}_u$. Thus

$$\bar{\Phi}_s(l) = \bar{E}\{e^{\pi j l n} e^{2\pi j l \sum_{i=0}^{n-1} \frac{x_i}{\Delta}}\}, \tag{6.5.11}$$

$$\bar{\Phi}_s^{(k)}(i,l) = e^{\pi j l k} \bar{E}\{e^{\pi j (i+l) n} e^{2\pi j (i+l) \sum_{m=0}^{n-1} \frac{x_m}{\Delta}} e^{2\pi j l \sum_{m=n}^{n-1+k} \frac{x_m}{\Delta}}\}. \tag{6.5.12}$$

Observe for later use that when $i = -l$, (6.5.12) simplifies to

$$\bar{\Phi}_s^{(k)}(-l,l) = e^{\pi j l k} \bar{E}\{e^{2\pi j l \sum_{m=n}^{n-1+k} \frac{x_m}{\Delta}}\}. \tag{6.5.13}$$

We now evaluate these expressions for the special cases of deterministic dc inputs.

Suppose that $x_k = x$ for all k, where $-b \le x < b$ is an irrational number. A dc input can be considered as an approximation to a slowly varying input, that is, to the case where the Sigma-Delta modulator has a large oversampling ratio. In this case we replace u_n by $s_n = n\beta$, where $\beta = (1/2 + x/\Delta)$, in (6.2.11)–(6.2.12) and evaluate

$$\bar{\Phi}_s(l) = \lim_{N\to\infty} \frac{1}{N} \sum_{n=1}^{N} e^{j2\pi l s_n} = \lim_{N\to\infty} \frac{1}{N} \sum_{n=1}^{N} e^{j2\pi l n\beta}$$

$$= \lim_{N\to\infty} \frac{1}{N} \sum_{n=1}^{N} e^{j2\pi l <n\beta>}. \tag{6.5.14}$$

From (6.3.8) this is

$$\bar{\Phi}_s(l) = \int_0^1 du e^{j2\pi u} = \begin{cases} 0; & l \ne 0 \\ 1; & l = 0 \end{cases} \tag{6.5.15}$$

Similarly,

$$\bar{\Phi}_s^{(k)}(i,l) = \lim_{N\to\infty} \frac{1}{N} \sum_{n=1}^{N} e^{j2\pi(i s_n + l s_{n+k})}$$

$$= \lim_{N\to\infty} \frac{1}{N} \sum_{n=1}^{N} e^{j2\pi(in\beta + l(n+k)\beta)} = e^{2\pi l k\beta} \lim_{N\to\infty} \frac{1}{N} \sum_{n=1}^{N} e^{j2\pi <n(i\beta + l\beta)>}$$

$$= \begin{cases} e^{2\pi lk\beta}; & i = -l \\ e^{2\pi lk\beta} \int_0^1 du e^{j2\pi u} = 0; & \text{otherwise,} \end{cases} \tag{6.5.16}$$

which is hence a special case of 6.5.13. Thus we have from (6.2.13)–(6.2.14) that

$$\bar{E}\{e_n\} = 0 \tag{6.5.17}$$

$$\bar{E}\{e_n^2\} = \frac{1}{12}, \tag{6.5.18}$$

which agrees with the uniform noise approximation, that is, these are exactly the time average moments one would expect with a sequence of uniform random variables. The second order properties, however, are quite different.

From (6.2.15) and 1.443.3 of [34]

$$R_e(k) = \sum_{l \neq 0} (\frac{1}{2\pi l})^2 e^{2\pi lk\beta} = \frac{1}{2}\frac{1}{\pi^2}\sum_{l=1}^{\infty} \frac{\cos(2\pi lk\beta)}{l^2}$$

$$= \frac{1}{12} - \frac{< k\beta >}{2}(1 - < k\beta >). \tag{6.5.19}$$

This does not correspond to a white noise process! The exponential expansion implies that the spectrum is purely discrete having amplitude

$$S_n = \begin{cases} 0; & \text{if } n = 0 \\ \frac{1}{(2\pi n)^2}; & \text{if } n \neq 0. \end{cases} \tag{6.5.20}$$

at frequencies $< n\beta > = < n(1/2 + x/\Delta) >$. Thus the locations and hence the amplitude of the quantizer error spectrum depend strongly on the value of the input signal. Thus as in the simple PCM case with a sinusoidal input, the Bennett white noise approximation inaccurately predicts the spectral nature of the quantizer noise process, which is neither continuous nor white.

A similar analysis can be carried out for the case of a sinuosoidal input. The interested reader is referred to [39] for the details.

6.6 Two-Stage Sigma-Delta Modulation

From the analysis for PCM and ordinary Sigma-Delta modulation, one might guess that the white noise approximation is never valid for quantizers with a small number of levels operating on simple deterministic signals. In this section we show that by cascading two Sigma-Delta modulators one can produce quantization noise that looks white in so far as its first and second order moments are concerned when the input is a fixed dc. Two-stage and multi-stage Sigma-Delta modulators were introduced by Uchimura and his colleagues [78], [63], [64]. The theoretical development here follows [82] and it can be extended using similar techniques to multistage Sigma-Delta modulators[18] as well as dithered multi-stage Sigma-Delta modulators[17].

A two-stage or cascaded Sigma-Delta modulator is shown in Figure 6.4.

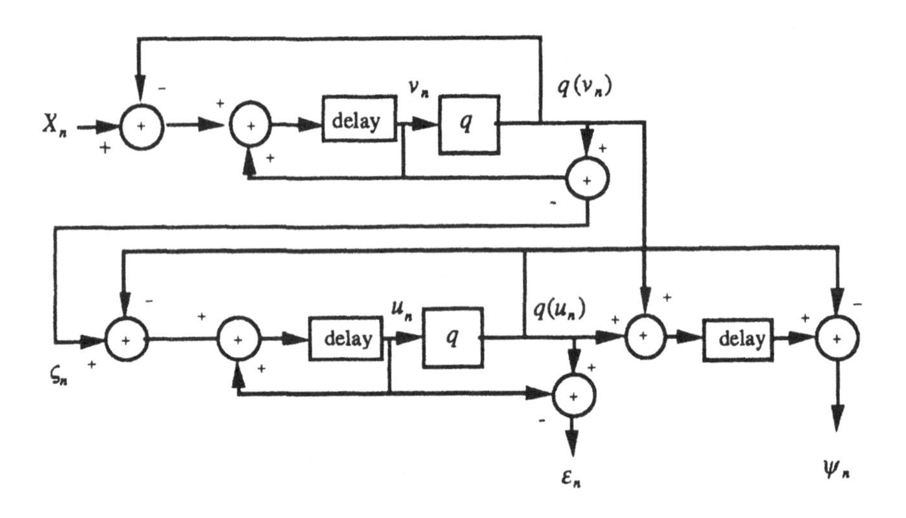

Figure 6.4: Two-Stage Sigma-Delta Modulator

Here the error sequence ζ_n from the first loop is the input to the second loop, where the quantizer error is denoted ϵ_n. From (6.5.1) the difference equations describing integrator states are

$$v_n = v_{n-1} + x_{n-1} - q(v_{n-1});\ n = 1, 2, \ldots \tag{6.6.1}$$

$$u_n = u_{n-1} + \zeta_{n-1} - q(u_{n-1});\ n = 1, 2, \ldots \tag{6.6.2}$$

The quantizers are exactly as those used with ordinary Sigma-Delta modulation, that is, uniform quantizers with M levels and bin width $\Delta = \Delta/(M-1)$, where the input x_n is between $-b$ and b. Note that this means that the first quantizer does not overload and hence ζ_n is in the range $[-b, b)$ and hence also the second quantizer does not overload. The output of the two stage sigma delta modulator is defined by

$$\psi_n = q(v_{n-1}) - q(u_n) + q(u_{n-1});\ n = 1, 2, \ldots, \tag{6.6.3}$$

a linear combination of the two quantizer outputs. As in (6.5.2) we have that

$$q(v_n) = x_{n-1} + \zeta_n - \zeta_{n-1},$$

$$q(u_n) = \zeta_{n-1} + \epsilon_n - \epsilon_{n-1}$$

and hence (6.6.3) becomes

$$\psi_n = x_{n-2} + \epsilon_n - 2\epsilon_{n-1} + \epsilon_{n-2}. \tag{6.6.4}$$

In contrast to (6.5.2), this has the interpretation of being the original signal plus a second order difference (discrete time second derivative) instead of the single order difference of the single loop system. Note that although this signal depends on the outputs of both stages, the first stage quantization noise cancels out and only the second stage noise remains. This fact is basic to the operation of the system. In particular, we need only know the behavior of the second stage quantization noise in order to find the behavior of the output. We shall see that the second stage noise is better behaved than the first stage noise.

Assuming that initially the integrator states (quantizer inputs) are $u_0 = v_0 = 0$, then applying (6.5.8) to both stages gives for $n = 1, 2, \ldots$

$$p_n = \frac{\psi_n}{\Delta} = \frac{1}{2} - < \frac{n}{2} + \sum_{k=0}^{n-1} \frac{x_k}{\Delta} >,$$

$$e_n = \frac{\epsilon_n}{\Delta} = \frac{1}{2} - < \frac{n}{2} + \sum_{k=0}^{n-1} p_k > = -\frac{1}{2} + < \sum_{k=0}^{n-1} \sum_{l=0}^{k-1} (\frac{1}{2} + \frac{x_l}{\Delta}) >$$

$$= -\frac{1}{2} + < \sum_{l=0}^{n-1} l(\frac{1}{2} + \frac{x_{n-l}}{\Delta}) > . \tag{6.6.5}$$

As in the ordinary Sigma-Delta modulator, we can modify (6.2.11)–(6.2.15) by replacing the quantizer input u_n by a sum term

$$s_n = \sum_{i=0}^{n-1} \sum_{l=0}^{i-1} (\frac{1}{2} + \frac{x_l}{\Delta}) = \sum_{l=0}^{n-1} l(\frac{1}{2} + \frac{x_{n-l}}{\Delta}). \tag{6.6.6}$$

and then proceed exactly as before. We here illustrate the results for the simple case of a dc input $x_n = x$. Define $\beta = 1/2 + x_n/\Delta$ as before and we have that

$$s_n = \frac{\beta}{2} n^2 - \frac{\beta}{2} n. \tag{6.6.7}$$

To evaluate the characteristic functions, we use the general version of Weyl's theorem [45]: If

$$c(n) = a_0 + a_1 n + a_2 n^2 + \cdots + a_k n^k$$

is a polynomial with real coefficients of degree $n \geq 1$ and if at least one of the coefficients a_i is irrational for $i \geq 1$, then for any Riemann integrable function g

$$\lim_{N\to\infty} \frac{1}{N} \sum_{n=1}^{N} g(< c(n) >) = \int_0^1 g(u)\, du. \tag{6.6.8}$$

We now use (6.6.8) exactly as (6.3.8) was used previously. We have with $c(n) = s_n$ in (6.6.8) that

$$\bar{\Phi}_s(l) = \lim_{N\to\infty} \frac{1}{N} \sum_{n=1}^{N} e^{2\pi j <c(n)>} = \int_0^1 e^{2\pi j u}\, du = \begin{cases} 0; & l \neq 0 \\ 1; & l = 0 \end{cases} \tag{6.6.9}$$

and

$$\bar{\Phi}_s^{(k)}(i,l) = \lim_{N\to\infty} \frac{1}{N} \sum_{n=1}^{N} e^{2\pi j <d(n)>}$$

where

$$d(n) = (i+l)\frac{\beta}{2}n^2 + (2il - i - l)\frac{\beta}{2}n + il(l-1)\frac{\beta}{2}$$

whence

$$\bar{\Phi}_s^{(k)}(i,l) = \begin{cases} 1; & i = -l = 0 \\ \int_0^1 e^{2\pi j u}\, du = 0; & \text{otherwise} \end{cases} \tag{6.6.10}$$

These characteristic functions are identical to those of the dithered PCM case in (6.4.6) and (6.4.7) and hence the conclusions are the same: the quantization noise is indeed white and its marginal first and second moments agree with those of a marginal distribution. Thus the Bennett model is a good approximation for the second stage quantization noise of a two-stage Sigma-Delta modulator driven by a dc (and only the second stage noise proves to be important in SQNR analysis).

The analysis can be extended to the case of a sinusoidal input, but the it is much more complicated and the noise is not white [82]. For third order and higher, however, the noise is white for sinusoids and finite sums of sinusoids[18].

Similar analyses can be applied to other Sigma-Delta architectures such as the multi-loop or higher order configurations (see He, Buzo, and Kuhlmann [47],[48]).

6.7 Delta Modulation

As a variation on the application of the uniform quantization noise analysis tools, we consider the important special case of a delta modulator. Recall that a delta modulator has the form of Figure 2.3 and, like the Sigma-Delta modulator, the Delta modulator can be used as an oversampled analog-to-digital converter. (In fact, Delta modulators were so used well before Sigma-Delta modulators and they have been widely used since.) A slight modification of the techniques used for Sigma-Delta modulation will yield the average distortion for the binary quantizer in a Delta modulator. Here, however, we also easily obtain the overall average distortion between the input signal and the final reproduction without much extra effort. The key conclusion is that for the simple test input signal considered, the quantization noise is well approximated by the uniform marginal approximation even though Bennett's conditions are violated. As with Sigma-Delta modulation, however, the noise is not at all white. These results are originally due to Iwersen [50]. Similar results using nonlinear difference equations were also obtained in unpublished work by T. J. Kim and D. L. Neuhoff.

Given a Delta modulator with an input signal u_n the output sequence y_n of the integrator is defined by the difference equation

$$y_n = y_{n-1} + u_n,$$

that is, it is the previous output plus the current input. We assume that the integrator has an initial *state* $u_0 = 0$ and that the above formula holds for all $n = 1, 2, \ldots$. Such an ideal integrator can be represented by a convolution by an IIR filter using the pulse response $h_k = 1$ for all nonnegative k and 0 otherwise. The filter is clearly not stable and hence care must be exercised. If the difference equation is changed slightly to

$$y_n = ay_{n-1} + u_n,$$

with $|a| < 1$, then the filter has pulse response $h_k = a^k$; $k = 0, 1, 2, \ldots$ and it is stable. In this case it is called a "leaky integrator." We focus on the ideal integrator. A delta modulator with a leaky or an ideal integrator is an example of *predictive quantization* [30].

It is more convenient to represent the filter as a first order autoregressive filter defined by the above difference equation. The encoder consists of a feedback loop containing the same integrator and a serial delay of one time unit. (In fact the delay represents the cumulative delay through the loop and points out the fact that the value subtracted from the current input depends on only previous inputs.)

The system can be completely described by difference equations and initial conditions. Define the following quantities as in the figure: Let $\tilde{x}_n$ denote the value subtracted from the input and

$$e_n = x_n - \tilde{x}_n = x_n - \hat{x}_{n-1} \tag{6.7.1}$$

the resulting difference that is put into the quantizer q defined by

$$q(e) = \begin{cases} +\Delta & e \geq 0 \\ -\Delta & e < 0 \, . \end{cases}$$

Δ is called the *step size* of the delta modulator. It is unfortunate that the same symbol is used here for a step size as was used in scalar quantization for the distance between quantization cell boundaries since, as will be seen, the distance between the corresponding boundaries for delta modulation turn out to be 2Δ instead of Δ. We accept this minor annoyance, however, so as to be consistent with the vast majority of the literature. The sequence of quantizer outputs $q(e_n)$ is put into the decoder integrator and the identical integrator in the encoder. The output of the integrator at time n is

$$\hat{x}_n = \hat{x}_{n-1} + q(e_n) \tag{6.7.2}$$

Because of the delay we have that $\tilde{x}_n = \hat{x}_{n-1}$. The system can be initialized by setting $\tilde{x}_0 = 0$ and hence $e_0 = x_0 - \tilde{x}_0 = x_0$. Note that in principle the possible reproduction values can eventually be anything of the form $\pm k\Delta$ for any integer k. At any time, however, only two values are possible given the past and hence the decoder rate is 1 bit/input symbol.

It should be pointed out that $\tilde{x}_n$ can be considered as a *prediction* of x_n based on the reproduction and hence e_n can be considered as a *prediction error* or *residual*. Define the binary quantizer error sequence as

$$\epsilon_n = q(e_n) - e_n . \tag{6.7.3}$$

Perhaps the most important mathematical property of a delta modulator is the following simple relation between the reconstruction error and the binary quantizer error:

$$\begin{aligned} \hat{x}_n - x_n &= \hat{x}_{n-1} + q(e_n) - x_n \\ &= q(e_n) - e_n = \epsilon_n . \end{aligned} \tag{6.7.4}$$

This points out that the overall mean squared error will be just as good as (and no better than) the mean squared value of the binary quantizer error sequence. Intuitively the error sequence should be less spread out and easier to quantize accurately with q than the original sequence would be. This means that solving for the moments and spectrum of the binary quantization noise immediately yields the overall analog-to-digital converter error moments, a property that did not hold for the Sigma-Delta modulators.

For later use we point out another important difference equation: Consider the $\{e_n\}$ sequence and observe that

$$e_n = x_n - \hat{x}_{n-1} = x_n - (\hat{x}_{n-2} + q(e_{n-1}))$$

$$= (x_n - x_{n-1}) + x_{n-1} - \hat{x}_{n-2} - q(e_{n-1})$$

$$= e_{n-1} + (x_n - x_{n-1}) - q(e_{n-1}) = (x_n - x_{n-1}) - \epsilon_{n-1}. \tag{6.7.5}$$

This provides a nonlinear difference equation or recursion for the input to the binary quantizer that points out the dependence on the difference or discrete-time derivative of the input, the previous binary quantizer input, and the previous binary quantizer output. Observe that a slight rearrangement provides the following relations among the reproduction error, quantizer error, and quantizer input sequence:

$$\hat{x}_n - x_n = \epsilon_n = (x_{n+1} - x_n) - e_{n+1}. \tag{6.7.6}$$

As in the Sigma-Delta derivations we need to require that the quantizer not overload. In Delta modulation this is equivalent to assuming that the delta modulator is *tracking* in the sense that $|x_n - \tilde{x}_n| \leq 2\Delta$, which implicitly assumes that the input is not changing too rapidly for the decoder to reproduce approximately. This condition may not be satisfied all the time, e.g., when the code is started at $n = 0$ a large constant input will cause a large prediction error. If the input does not vary rapidly, however, the code will eventually provide a reproduction that is close to the input. This should be the case most of the time if the oversampling ratio is high.

As with Sigma-Delta modulation, the traditional means of analysis of the average quantizer error in a delta modulation scheme is to replace the quantizer noise sequence $\{\epsilon_n\}$ by an additive memoryless random process with a uniform distribution on its input range. If the delta modulator is tracking, this input range is $[-\Delta, \Delta)$ and in this case one easily finds from (6.7.4) that

$$E[(\hat{x}_n - x_n)^2] = E[\epsilon_n^2] = \frac{\Delta^2}{3}. \tag{6.7.7}$$

(Recall that the symbol Δ here corresponds to 2Δ in the scalar case and hence the above is just our old friend $\Delta^2/12$ with a redefined Δ.) As in the Sigma-Delta case, however, the conditions under which the white noise assumption is valid do not hold.

The difference equations and initial conditions describing the simple delta modulator can be written as

$$e_0 = x_0;\ \epsilon_0 = q(x_0) - x_0, \tag{6.7.8}$$

$$e_n = e_{n-1} + (x_n - x_{n-1}) - q(e_{n-1}) = (x_n - x_{n-1}) - \epsilon_{n-1}, \tag{6.7.9}$$

$$\epsilon_n = q(e_n) - e_n. \tag{6.7.10}$$

These relations hold for arbitrary input signals. In order to proceed, we now make assumptions about the input sequence. Roughly speaking, we assume that the signal is not too big at time 0 (when the machine is turned on) and that it cannot change too much in a single time unit. In particular, if the output of the binary quantizer is $\pm\Delta$, then

$$|x_0| < 2\Delta, \tag{6.7.11}$$

$$|x_n - x_{n-1}| < \Delta. \tag{6.7.12}$$

These assumptions are required for the subsequent analysis. The second assumption reflects the oversampling assumption: If an original continuous time waveform is sampled at many times the Nyquist rate, then the sample values should not change much in adjacent samples. The first assumption can be viewed as stating that the delta modulator is not too far off when turned on, e.g., that it has perhaps already been tracking the input and provides a reasonable approximation. Note that the signal does not have a uniform bound over time, but the above assumptions require that it not grow too rapidly, e.g., from the triangle inequality we must have that

$$|x_n| = |\sum_{k=1}^{n}(x_k - x_{k-1}) + x_0|$$

$$\leq \sum_{k=1}^{n} |x_k - x_{k-1})| + |x_0| < n\Delta + 2\Delta. \tag{6.7.13}$$

In the common terminology for delta modulation, we consider signals for which there is only granular noise and no slope overload noise.

We begin with a simple property showing that the quantization error sequence and the prediction error sequence are both bounded.

Theorem 6.7.1 *For $n = 0, 1, \ldots$*

$$-2\Delta < e_n < 2\Delta \tag{6.7.14}$$

$$-\Delta < \epsilon_n < \Delta. \tag{6.7.15}$$

Proof:

First observe that (6.7.14) immediately implies (6.7.15) since if the input to a binary quantizer with output levels $\pm\Delta$ has maximum range $[-2\Delta, 2\Delta]$, then the maximum quantization error magnitude is Δ. To prove (6.7.14) use induction. For $n = 0$ we have $e_0 = x_0$ which meets the condition by assumption. Assume that (6.7.14) (and hence (6.7.15)) holds for a fixed n and hence

$$|\epsilon_{n-1}| = |x_{n-1} - \hat{x}_{n-1}| < \Delta.$$

For $n+1$ we then have using the triangle inequality

$$|e_n| = |x_n - \hat{x}_{n-1}| = |x_n - \hat{x}_{n-1} + \hat{x}_{n-1} - \hat{x}_{n-1}|$$

$$\leq |x_n - \hat{x}_{n-1}| + |\hat{x}_{n-1} - \hat{x}_{n-1}| < 2\Delta,$$

proving the claim.

We are now ready to solve the nonlinear difference equations of (6.7.8)–(6.7.10). As a first step we combine (6.7.8)–(6.7.10) into a single difference equation for the binary quantization error ϵ_n: For $n = 1, 2, \ldots$

$$\epsilon_n = q(x_n - x_{n-1} - \epsilon_{n-1}) - (x_n - x_{n-1} - \epsilon_{n-1}). \tag{6.7.16}$$

Define the indicator function (the so-called *unit step function*)

$$\chi(r) = \begin{cases} 1 & \text{if } r \geq 0 \\ 0 & \text{if } r < 0 \end{cases}$$

Then (6.7.16) becomes

$$\epsilon_n = 2\Delta\chi(x_n - x_{n-1} - \epsilon_{n-1}) - \Delta - (x_n - x_{n-1} - \epsilon_{n-1}). \tag{6.7.17}$$

The following theorem provides the solution to this nonlinear difference equation with the initial condition of (6.7.8).

Theorem 6.7.2 *For* $n = 1, 2, \ldots$

$$\frac{\epsilon_n}{2\Delta} = \frac{1}{2} - < \frac{x_n - x_{n-1}}{2\Delta} - \frac{\epsilon_{n-1}}{2\Delta} > . \tag{6.7.18}$$

For all $n = 0, 1, \ldots$

$$\frac{\epsilon_n}{2\Delta} = \frac{1}{2} - < \frac{x_n}{2\Delta} + \frac{1-(-1)^n}{4} >$$

$$= \frac{1}{2} - < \frac{x_n}{2\Delta} + \frac{n}{2} > = \begin{cases} \frac{1}{2} - < \frac{x_n}{2\Delta} > & n \text{ even} \\ \frac{1}{2} - < \frac{x_n}{2\Delta} + \frac{1}{2} > & n \text{ odd}. \end{cases} \tag{6.7.19}$$

Comments:

Eq. (6.7.18) provides an alternative representation of the recursion in terms of the fractional operator and is a key step in proving (6.7.19), the principal result of the theorem and of this section and the key to finding the sample moments. Analogous to the Sigma-Delta analysis, (6.7.19) shows that when the input signals meet the assumed constraints, the quantization error is a simple memoryless nonlinear function of the current input.

Proof:

Abbreviate $(x_n - x_{n-1} - \epsilon_{n-1})/2\Delta$ by γ_n so that (6.7.17) becomes

$$\frac{\epsilon_n}{2\Delta} = \chi(\gamma_n) - \gamma_n - \frac{1}{2}$$

which we write as

$$\frac{\epsilon_n}{2\Delta} = \frac{1}{2} - [\gamma_n + 1 - \chi(\gamma_n)]. \qquad (6.7.20)$$

Consider the leftmost term in the brackets:

$$\gamma_n + 1 = \frac{x_n - x_{n-1} - \epsilon_{n-1}}{2\Delta} + 1 = \frac{e_n}{2\Delta} + 1.$$

Theorem 6.7.1 then implies that

$$0 < \gamma_n + 1 < 2$$

and hence

$$< \gamma_n + 1 >= \begin{cases} \gamma_n + 1 & \text{if } \gamma_n + 1 < 1 \\ \gamma_n + 1 - 1 = \gamma_n & \text{if } \gamma_n + 1 \geq 1 \end{cases};$$

that is,

$$< \gamma_n + 1 >= \gamma_n - \chi(\gamma_n)$$

which with (6.7.20) proves (6.7.18). Induction is used to prove (6.7.19). The result for $n = 0$ follows by direct substitution (0 is considered even in this result). Assume the result holds for a fixed n and hence

$$\frac{\epsilon_n}{2\Delta} = \begin{cases} \frac{1}{2} - < \frac{x_n}{2\Delta} > & \text{if } n \text{ even} \\ \frac{1}{2} - < \frac{x_n}{2\Delta} + \frac{1}{2} > & \text{if } n \text{ odd} \end{cases}$$

$$= \begin{cases} \frac{1}{2} - \frac{x_n}{2\Delta} + \lfloor \frac{x_n}{2\Delta} \rfloor & \text{if } n \text{ even} \\ -\frac{x_n}{2\Delta} + \lfloor \frac{x_n}{2\Delta} \rfloor & \text{if } n \text{ odd} \end{cases} \qquad (6.7.21)$$

Using (6.7.4) and (6.7.21) yields

$$\frac{\hat{x}_n}{2\Delta} = \begin{cases} \frac{1}{2} + \lfloor \frac{x_n}{2\Delta} \rfloor & \text{if } n \text{ even} \\ \lfloor \frac{x_n}{2\Delta} + \frac{1}{2} \rfloor & \text{if } n \text{ odd} \end{cases} \tag{6.7.22}$$

and hence using (6.7.18)

$$\epsilon_{n+1} = \frac{1}{2} - < \frac{x_{n+1}}{2\Delta} - \frac{\hat{x}_n}{2\Delta} >$$

$$= \begin{cases} \frac{1}{2} - < \frac{x_{n+1}}{2\Delta} - \frac{1}{2} - \lfloor \frac{x_n}{2\Delta} \rfloor > & \text{if } n \text{ even} \\ \frac{1}{2} - < \frac{x_{n+1}}{2\Delta} - \lfloor \frac{x_n}{2\Delta} + \frac{1}{2} \rfloor > & \text{if } n \text{ odd}\ . \end{cases}$$

Since $n + 1$ is odd (even) if n is even (odd) and since integers inside the fractional operator can be removed without affecting its value, this becomes

$$\epsilon_{n+1} = \begin{cases} \frac{1}{2} - < \frac{x_{n+1}}{2\Delta} + \frac{1}{2} > & \text{if } n+1 \text{ odd} \\ \frac{1}{2} - < \frac{x_{n+1}}{2\Delta} > & \text{if } n+1 \text{ even}\ , \end{cases} \tag{6.7.23}$$

proving the theorem by induction.

We now apply the solution to the difference equation to the evaluation of the moments of the binary quantizer noise (and hence also of the overall analog-to-digital conversion noise).

As before, ϵ_n is given by

$$\epsilon_n = q(e_n) - e_n = \hat{x}_n - x_n.$$

and we wish to evaluate the following moments:

$$M\{\epsilon_n\} = \lim_{N\to\infty} \frac{1}{N} \sum_{n=0}^{N-1} \epsilon_n,$$

$$M\{\epsilon_n^2\} = \lim_{N\to\infty} \frac{1}{N} \sum_{n=0}^{N-1} \epsilon_n^2,$$

and

$$r_\epsilon(k) = \lim_{N\to\infty} \frac{1}{N} \sum_{n=0}^{N-1} \epsilon_n \epsilon_{n+k}.$$

To evaluate these moments we focus on an input of the form

$$x_n = bn + a, \tag{6.7.24}$$

for fixed constants a and b, that is, a straight line input. In order to meet the required constraints of (6.7.8) (and thereby avoid slope overload) we require that

$$|a| \leq 2\Delta; \ |b| \leq \Delta. \tag{6.7.25}$$

This input is a good model for the situation where x_n varies slowly. The principal application in mind was mentioned in the first section: Suppose that x_n is formed by *oversampling* a continuous waveform, that is, by sampling at many times the Nyquist rate. In this case x_n will vary slowly and will look like straight lines for long periods of time.

In this example Theorem 6.7.2 implies that

$$\epsilon_n = \frac{1}{2} - < a + bn + \frac{n}{2} > = \frac{1}{2} - < a + (\frac{1}{2} + b)n >$$

and therefore if we define $\beta = 1/2 + b$

$$M\{\epsilon_n\} = \frac{1}{2} - \lim_{N\to\infty} \frac{1}{N} \sum_{n=0}^{N-1} < \beta n + a > . \tag{6.7.26}$$

and

$$M\{\epsilon_n^2\} = \frac{1}{4} - \frac{1}{2} \lim_{N\to\infty} \frac{1}{N} \sum_{n=0}^{N-1} < \beta n + a >$$

$$+ \lim_{N\to\infty} \frac{1}{N} \sum_{n=0}^{N-1} < \beta n + a >^2 . \tag{6.7.27}$$

We consider two cases. If β is an irrational number, then we can invoke Weyl's theorem to conclude

$$\lim_{n\to\infty} \frac{1}{n} \sum_{k=0}^{n-1} \epsilon_n = 0, \tag{6.7.28}$$

and hence the sample mean of the error is indeed 0. Similarly with $f(r) = r^2$ in (6.7.26) we have

$$\lim_{N\to\infty} \frac{1}{N} \sum_{n=0}^{N-1} \epsilon_n^2 = \frac{\Delta^2}{3}, \tag{6.7.29}$$

and hence the time average energy of the error sequence is also consistent with the uniform marginal distribution assumption! Since this is also the overall analog-to-digital conversion average distortion, this result supports the use of the common white noise approximation even though the conditions for that approximation are violated!

To compare this result with the results for ordinary scalar uniform quantization, observe that if we choose a scalar quantizer with a range of B, $M = 2^R$ levels, where R is the rate, and with bin width $\Delta_s = B/2^R$, the distance between the reproduction levels, equal to the bin width of the delta modulator, 2Δ, then both schemes yield an average mean squared error of

$$\frac{\Delta_s^2}{12}.$$

The delta modulator, however, has rate 1 and the scalar quantizer must have rate $R = \log(B/\Delta)$, which can be much larger.

For completeness we show how to modify the analysis for a rational β. If β is a rational number, then it will have the form $\beta = K/N$ in lowest terms, that is, K and N share no common divisors other than 1. In this case

$$\epsilon_n = \frac{1}{2} - < a + \frac{K}{N} n > \tag{6.7.30}$$

is a periodic sequence with period N and hence

$$M\{\epsilon_n\} = \frac{1}{N} \sum_{n=0}^{N-1} \epsilon_n$$

$$= \frac{1}{2} - \frac{1}{N} \sum_{n=0}^{N-1} < a + \frac{K}{N} n > \tag{6.7.31}$$

and

$$M\{\epsilon_n^2\} = \frac{1}{N} \sum_{n=0}^{N-1} \epsilon_n^2$$

$$= \frac{1}{4} - 2\frac{1}{N} \sum_{n=0}^{N-1} < a + \frac{K}{N} n > + \frac{1}{N} \sum_{n=0}^{N-1} < a + \frac{K}{N} n >^2 . \tag{6.7.32}$$

Unlike the case of irrational β, these sums strongly depend on the value of a. Hence for convenience and to provide a comparison between the rational and irrational cases, we assume here that $a = 0$. In this case the sums can easily be evaluated by noting that the collection of values $< nK/N >$, $n = 0, 1, \ldots, N-1$ is just a reordered version of the values $< n/N >$ and hence

$$\frac{1}{N} \sum_{n=0}^{N-1} < \frac{K}{N} n > = \frac{1}{N} \sum_{n=0}^{N-1} \frac{k}{N} = \frac{1}{2} \frac{N-1}{N}$$

and

$$\frac{1}{N}\sum_{n=0}^{N-1} < \frac{K}{N}n >^2 = \frac{1}{N}\sum_{n=0}^{N-1}(\frac{k}{N})^2 = \frac{1}{3} - \frac{1}{2N} + \frac{1}{6N^2}$$

and therefore

$$M\{\epsilon_n\} = \frac{1}{N} \tag{6.7.33}$$

$$M\{\epsilon_n^2\} = \frac{1}{12} + \frac{1}{6N^2}. \tag{6.7.34}$$

We have used the standard formulas for sums of the form $\sum_{k=1}^{n} k$ and $\sum_{k=1}^{n} k^2$ in the above evaluation. Observe that these moments are not the same as in the irrational case since there is an additional term depending on the period. These results are asymptotically consistent, however, since if we take a sequence of rational numbers converging to an irrational number, then the period goes to infinity and the moments for the rational numbers converge to those of the limiting irrational number. Thus, for example, simulating randomly chosen continuous numbers on a digital computer will with high probability produce a rational number with a large denominator and a long period. Thus the irrational result provides a good approximation for such rational numbers. It should be emphasized, however, that for any given rational β, the sample moments of the error are *not* exactly consistent with the common uniform assumption. It is only in the limit of closely approximating an irrational number that the uniform assumption is good.

Lastly we turn to the evaluation of the autocorrelation function of the quantizer noise.

From Theorem 6.7.2 we can write the binary quantizer error sequence as

$$\epsilon_n = g(\frac{n}{2} + x_n), \tag{6.7.35}$$

where

$$g(r) = \frac{1}{2} - < r > \tag{6.7.36}$$

is a memoryless nonlinearity. As in the Sigma-Delta analysis $g(r)$ is a periodic function with period 1 and hence has a Fourier series expansion

$$g(x) = \sum_{l=-\infty}^{\infty} \hat{g}(l)e^{2\pi jlx} \tag{6.7.37}$$

where

$$\hat{g}(l) = \int_0^1 dx g(x)e^{-2\pi jlx}. \tag{6.7.38}$$

In the case $g(x) = 1/2- < x >$ this is

$$\begin{cases} 0 & \text{if } l = 0 \\ -\frac{j}{2\pi l} & \text{if } l \neq 0. \end{cases} \qquad (6.7.39)$$

Using this expansion for ϵ_n and defining $s_n = x_n + n/2$ we have

$$r_\epsilon(k) = M\{\epsilon_n\epsilon_{n+k}\} = M\{\left(\sum_i \hat{g}(i)e^{j2\pi i s_n}\right)\left(\sum_l \hat{g}(l)e^{j2\pi l s_{n+k}}\right)\}$$

$$= \sum_i \sum_l \hat{g}(i)\hat{g}(l)M\{e^{j2\pi(is_n + ls_{n+k})}\}, \qquad (6.7.40)$$

if we assume as before that the various limits exist and can be interchanged. This can be written in terms of the the sample joint characteristic function

$$\Phi_s^{(k)}(i,l) = M\{e^{j2\pi(is_n+ls_{n+k})}\} = \lim_{N\to\infty} \frac{1}{N}\sum_{n=1}^{N} e^{j2\pi(is_n+ls_{n+k})}, \qquad (6.7.41)$$

as

$$r_\epsilon(k) = \sum_i \sum_l \hat{g}(i)\hat{g}(l)\Phi_s^{(k)}(i,l). \qquad (6.7.42)$$

To find the autocorrelation, we need to evaluate the joint characteristic function $\Phi_s^{(k)}(i,l)$. This we now do for the special case of x_n as in (6.7.24) with b (and hence $\beta = b + 1/2$) irrational. We have that $s_n = a + \beta n$

$$\Phi_s^{(k)}(i,l) = \lim_{N\to\infty} \frac{1}{N}\sum_{n=1}^{N} e^{j2\pi(i(a+\beta n)+l(a+\beta(n+k))}$$

$$= e^{j2\pi lk\beta} \lim_{N\to\infty} \frac{1}{N}\sum_{n=1}^{N} e^{j2\pi((i+l)(a+\beta n))}$$

$$= e^{j2\pi(lk\beta} \lim_{N\to\infty} \frac{1}{N}\sum_{n=1}^{N} e^{j2\pi(i+l)<a+\beta n>}. \qquad (6.7.43)$$

Again using (6.7.29) this becomes

$$\Phi_s^{(k)}(i,l) = e^{j2\pi[lk\beta]} \int_0^1 d\theta e^{j2\pi(i+l)\theta}$$

$$= \begin{cases} e^{j2\pi lk\beta} & \text{if } i = -l \\ 0 & \text{if } i + l \neq 0 \end{cases} \qquad (6.7.44)$$

and hence

$$r_\epsilon(k) = \sum_l \hat{g}(l)\hat{g}(-l)e^{j2\pi lk\beta} = \sum_l |\hat{g}(l)|^2 e^{j2\pi lk\beta}$$

$$= \sum_{l\neq 0} \left(\frac{1}{2\pi l}\right)^2 e^{j2\pi lk\beta}. \tag{6.7.45}$$

The above formula (which is originally due to Iwersen [50]) has the form

$$r_\epsilon(k) = \sum_{l=-\infty}^{\infty} S_l e^{2\pi kl\beta}, \tag{6.7.46}$$

which is a generalized Fourier series (or Bohr-Fourier series) with spectral coefficients

$$S_l = \frac{1}{(2\pi l)^2},\ l \neq 0 \tag{6.7.47}$$

which is exactly the same as those found for Sigma-Delta modulation with a dc input of (6.5.20). A moment's reflection should show that this is to be expected: A Sigma Delta modulator can be considered as a Delta modulator with an integrator in front of it. Hence putting a dc signal into a Sigma Delta modulator is equivalent to putting an integrated dc, a ramp, into a Delta modulator. Hence either analysis could have been used to arrive at the other. The cases were handled separately, however, to demonstrate the tools. It should be emphasized, however, that there is an important difference between the two cases. In the Delta modulation case one immediately has the moments and spectrum of the overall analog-to-digital conversion noise–they arre the same as the binary quantizer noise moments and spectrum. In the Sigma-Delta case these must be derived.

To summarize the development for Delta modulation: The quantizer and overall noises are not white, but the marginal moments of both agree with that predicted by the white noise approximation, that is,

$$M\{(x_n - \hat{x}_n)^2\} = M\{\epsilon_n^2\} = \frac{\Delta^2}{3}. \tag{6.7.48}$$

Exercises

1. Work through the example described at the beginning of Section 6.4, that is, a sinusoid with a random phase.

2. Work through the second example described in Section 6.4, that is, show that a iid uniform process into a uniform quantizer (with a correctly chosen range) yields an iid uniform error sequence.

3. Consider the example of PCM with a dithering signal of Section 6.4, but now suppose that the input u_n is a constant, say $u_n = u$. This might be a reasonable approximation if the input is oversampled and hence changes very slowly. Assume that the dithering signal w_n is an iid white process is in Section 6.4. The decoder operates as follows: For a fixed (large) N define

$$\hat{u}_N = \frac{1}{N} \sum_{n=0}^{N-1} q(u + w_n).$$

Intuitively, the decoder will form reproduction levels at the original sampling rate by summing up the previous N quantizer outputs and scaling. Show that with probability one

$$\lim_{N \to \infty} |\hat{u}_N - u| = 0,$$

that is, dithering can actually decrease the quantizer error for this special case of quantizing an unknown dc signal.

4. Prove (6.5.8) using induction.

5. Suppose that a delta modulator has a leaky integrator with pulse response $h_k = a^k$, $k \geq 0$. (The same integrator is used in both encoder and decoder.) Does (6.7.4) still hold? What is the analogous difference equation to (6.7.5)?

6. Derive the power spectrum of the Delta modulator quantization noise given by (6.7.30), that is, for the case of a ramp input with a rational slope.

Bibliography

[1] P. Algoet and T. Cover. A sandwitch proof of the Shannon-McMillan-Breiman theorem. *Annals of Probability*, 16:899–909, 1988.

[2] D. S. Arnstein. Quantization error in predictive coders. *IEEE Transactions on Communications*, COM-23:423–429, April 1975.

[3] A. R. Barron. The strong ergodic theorem for densities: generalized Shannon-McMillan-Breiman theorem. *Ann. Probab.*, 13:1292–1303, 1985.

[4] W. R. Bennett. Spectra of quantized signals. *Bell Systems Technical Journal*, 27:446–472, July 1948.

[5] T. Berger. *Rate Distortion Theory*. Prentice-Hall Inc., Englewood Cliffs, New Jersey, 1971.

[6] R. E. Blahut. Computation of channel capacity and rate-distortion functions. *IEEE Transactions on Information Theory*, IT-18:460–473, 1972.

[7] R. E. Blahut. *Principles and Practice of Information Theory*. Addison-Wesley, Reading, Mass., 1987.

[8] S. Bochner. Beitrage zur theorie der fastperiodischen funktionen I, II. *Math. Ann.*, 96:119–147, 1927.

[9] S. Bochner and J. von Neumann. Almost periodic functions in groups, II. *Trans. Amer. Math. Soc.*, 37:21–50, 1935.

[10] H. Bohr. *Almost Periodic Functions*. Chelsea, New York, 1947. Translation by Harvey Cohn.

[11] J. A. Bucklew and G. L. Wise. Multidimensional asymptotic quantization theory with rth power distortion measures. *IEEE Transactions on Information Theory*, IT-28:239–247, March 1982.

[12] J. C. Candy. A use of limit cycle oscillations to obtain robust analog-to-digital converters. *IEEE Transactions on Communications*, COM-22:298–305, March 1974.

[13] J. C. Candy. A use of double integration in sigma delta modulation. *IEEE Transactions on Communications*, COM-33:249–258, March 1985.

[14] J. C. Candy. Decimation for sigma delta modulation. *IEEE Transactions on Communications*, COM-34:72–76, January 1986.

[15] J. C. Candy and O. J. Benjamin. The structure of quantization noise from sigma-delta modulation. *IEEE Transactions on Communications*, COM-29:1316–1323, Sept. 1981.

[16] J. C. Candy, Y. C. Ching, and D. S. Alexander. Using triangularly weighted interpolation to get 13-bit PCM from a sigma delta modulator. *IEEE Transactions on Communications*, pages 1268–1275, November 1976.

[17] W. Chou and R. M. Gray. Dithering and its effects on multi-stage sigma-delta modulation. Submitted for possible publication, 1989.

[18] W. Chou, P. W. Wong, and R. M. Gray. Multi-stage sigma-delta modulation. *IEEE Transactions on Information Theory*, 1989. To appear.

[19] T. A. C. M. Claasen and A. Jongepier. Model for the power spectral density of quantization noise. *IEEE Trans. on ASSP*, ASSP-29:914–917, August 1981.

[20] A. G. Clavier, P. F. Panter, and D. D. Grieg. Distortion in a pulse count modulation system. *AIEE Transactions*, 66:989–1005, 1947.

[21] A. G. Clavier, P. F. Panter, and D. D. Grieg. PCM distortion analysis. *Electrical Engineering*, pages 1110–1122, November 1947.

[22] J. H. Conway and N. J. A. Sloane. *Sphere Packings, Lattices and Groups*. Springer-Verlag, New York, 1988.

[23] I. Csiszàr. On an extremum problem of information theory. *Studia Scientiarum Mathematicarum Hungarica*, pages 57–70, 1974.

[24] I. Csiszàr and G. Tusnady. Information geometry and alternating minimization procedures. *Statistics and Decisions*, pages 205–237, 1984. Supplement Issue No. 1.

[25] W. B. Davenport and W. L Root. *An Introduction to the Theory of Random Signals and Noise*. McGraw-Hill, New York, 1958.

[26] J. G. Dunham. A note on the abstract alphabet block source coding with a fidelity criterion theorem. *IEEE Transactions on Information Theory*, IT-24:760, November 1978.

[27] G. D. Forney. The Viterbi algorithm. *Proceedings IEEE*, 61:268–278, March 1973.

[28] R. G. Gallager. *Information theory and reliable communication*. John Wiley & Sons, NY, 1968.

[29] A. Gersho. Asymptotically optimal block quantization. *IEEE Transactions on Information Theory*, IT-25:373–380, July 1979.

[30] A. Gersho and R. M. Gray. *Vector Quantization and Signal Compression*. Kluwer Academic Press, 1990.

[31] J. D. Gibson and K. Sayood. Lattice quantization. *Advances in electronics and electron physics*, 72, 1988.

[32] H. Gish and J. N. Pierce. Asymptotically efficient quantizing. *IEEE Transactions on Information Theory*, IT-14:676–683, September 1968.

[33] T. J. Goblick and J. L. Holsinger. Analog source digitization: a comparison of theory and practice. *IEEE Transactions on Information Theory*, IT-13:323–326, 1967.

[34] I. S. Gradshteyn and I. M. Ryzhik. *Table of Integrals, Series, and Products*. Academic Press, New York, 1965.

[35] R. M. Gray. Information rates of autoregressive processes. *IEEE Transactions on Information Theory*, IT-16:412–421, 1970.

[36] R. M. Gray. Toeplitz and circulant matrices: II. Technical report, Stanford University, April 1977. (Available on request from author.).

[37] R. M. Gray. *Probability, Random Processes, and Ergodic Properties*. Springer-Verlag, New York, 1988.

[38] R. M. Gray. *Mathematical Information Theory*. Springer-Verlag, New York, 1990.

[39] R. M. Gray, W. Chou, and P. W. Wong. Quantization noise in single-loop sigma-delta modulation with sinusoidal inputs. *IEEE Transactions on Communications*, 1989. To Appear.

[40] R. M. Gray and L. D. Davisson. *Random Processes: A Mathematical Approach for Engineers.* Prentice-Hall, Englewood Cliffs, New Jersey, 1986.

[41] R. M. Gray, M. O. Dunham, and R. Gobbi. Ergodicity of Markov channels. *IEEE Transactions on Information Theory*, IT-33:656–664, September 1987.

[42] R. M. Gray and J. C. Kieffer. Asymptotically mean stationary measures. *Annals of Probability*, 8:962–973, Oct. 1980.

[43] R. M. Gray, D. L. Neuhoff, and J. K. Omura. Process definitions of distortion rate functions and source coding theorems. *IEEE Trans. on Info. Theory*, IT-21:524–532, 1975.

[44] U. Grenander and G. Szego. *Toeplitz Forms and Their Applications.* University of California Press, Berkeley and Los Angleles, 1958.

[45] F. J. Hahn. On affine transformations of compact abelian groups. *American Journal of Mathematics*, 85:428–446, 1963. Errata in Vol. 86, pp. 463–464, 1964.

[46] G. H. Hardy, J. E. Littlewood, and G. Polya. *Inequalities.* Cambridge Univ. Press, London, 1952. Second Edition, 1959.

[47] N. He, A. Buzo, and F. Kuhlmann. A frequency domain waveform speech compression system based on product vector quantizers. In *Proceedings of ICASSP*, Tokyo, Japan, April 1986.

[48] N. He, A. Buzo, and F. Kuhlmann. Multi-loop sigma-delta quantization, 1987. Submitted for possible publication.

[49] H. Inose and Y. Yasuda. A unity bit coding method by negative feedback. In *Proceedings of the IEEE*, volume 51, pages 1524–1535, November 1963.

[50] J. E. Iwersen. Calculated quantizing noise of single-integration delta-modulation coders. *BSTJ*, pages 2359–2389, September 1969.

[51] R. C. Jaffee. *Casual and statistical analysis of dithered systems containing three level quantizer.* M.S. thesis, M.I.T., 1959.

[52] N. S. Jayant and P. Noll. *Digital Coding of Waveforms.* Prentice-Hall, Englewood Cliffs,New Jersey, 1984.

[53] F. Jelinek and J. B. Anderson. Instrumentable tree encoding of information sources. *IEEE Trans. Info. Theory*, IT-17:118–119, Jan 1971.

[54] J. C. Kieffer. A generalization of the pursley-davisson-mackenthun universal variable-rate coding theorem. *IEEE Trans. on Information Theory*, IT-23:694–697, 1977.

[55] J. C. Kieffer. Sliding-block coding for weakly continuous channels. *IEEE Trans. Inform. Theory*, IT-28:2–10, 1982.

[56] J. C. Kieffer. *Elementary and Advanced Information Theory.* 1989. Book in Progress.

[57] J. C. Kieffer and J. G. Dunham. On a type of stochastic stability for a class of encoding schemes. *IEEE Transactions on Information Theory*, IT-29:793–797, 1983.

[58] J. C. Kieffer and M. Rahe. Markov channels are asymptotically mean stationary. *SIAM Journal of Mathematical Analysis*, 12:293–305, 1980.

[59] Yu. N. Linkov. Evaluation of epsilon entropy of random variables for small epsilon. *Problems of Information Transmission*, 1:12–18, 1965. Translated from Problemy Peredachi Informatsii, Vol. 1, 18–26.

[60] L. Ljung. *System Identification.* Prentice-Hall, Englewood Cliffs, NJ, 1987.

[61] S. P. Lloyd. Least squares quantization in PCM. Unpublished technical note, Bell Laboratories, 1957. Portions presented at the Institute of Mathematical Statistics Meeting, Atlantic City, New Jersey, September 1957. Published in the March 1982 special issue on quantization of the IEEE Transactions on Information Theory.

[62] K. Marton. On the rate distortion function of stationary sources. *Problems of Control and Information Theory*, 4:289–297, 1975.

[63] Y. Matsuya, K. Uchimura, A. Iwata, T. Kobayashi, and M. Ishikawa. A 16b oversampling conversion technology using triple integration noise shaping. In *Proceedings 1987 IEEE International Solid-State Circuits Conference*, pages 48–49, February 1987.

[64] Y. Matsuya, K. Uchimura, A. Iwata, T. Kobayashi, M. Ishikawa, and T. Yoshitome. A 16-bit oversampling A-to-D conversion technology using triple-integration noise shaping. *IEEE Journal of Solid-state Circuits*, SC-22:921–929, December 1987.

[65] J. Max. Quantizing for minimum distortion. *IEEE Transactions on Information Theory*, pages 7–12, March 1960.

[66] R. McEliece. *The Theory of Information and Coding.* Cambridge University Press, New York, NY, 1984.

[67] D. L. Neuhoff. Source coding strategies: simple quantizers vs. simple noiseless codes. In *Proceedings 1986 Conf. on Information Sciences and Systems*, volume 1, pages 267–271, March 1986.

[68] K. Petersen. *Ergodic Theory.* Cambridge University Press, Cambridge, 1983.

[69] M. S. Pinsker. *Information and information stability of random variables and processes.* Holden Day, San Francisco, 1964. Translated by A. Feinstein from the Russian edition published in 1960 by Izd. Akad. Nauk. SSSR.

[70] S. O. Rice. Mathematical analysis of random noise. In N. Wax, editor, *Selected papers on noise and stochastic processes*, pages 133–294. Dover, New York, NY, 1954. Reprinted from Bell Systems Technical Journal, Vol. 23:282–332 (1944) and Vol. 24: 46–156 (1945).

[71] L. Schuchman. Dither signals and their effects on quantization noise. *IEEE Transactions on Communication Technology*, COM-12:162–165, December 1964.

[72] M. P. Schutzenberger. On the quantization of finite dimensional messages. *Inform. Control*, 1:153–158, 1958.

[73] C. E. Shannon. A mathematical theory of communication. *Bell Systems Technical Journal*, 27:379–423,623–656, 1948.

[74] C. E. Shannon. Coding theorems for a discrete source with a fidelity criterion. *IRE National Convention Record, Part 4*, pages 142–163, 1959.

[75] D. Slepian. On delta modulation. *Bell System Technical Journal*, 51:2101–2136, 1972.

[76] A. B. Sripad and D. L. Snyder. A necessary and sufficient condition for quantization errors to be uniform and white. *IEEE Trans. on ASSP*, ASSP-25:442–448, October 1977.

[77] J. Storer. *Data Compression.* Computer Science Press, Rockville, Maryland, 1988.

[78] K. Uchimura, T. Hayashi, T. Kimura, and A. Iwata. VLSI-A to D and D to A converters with multi-stage noise shaping modulators. In *Proceedings 1986 ICASSP*, pages 1545–1548, Tokyo, 1986.

[79] A. J. Viterbi and J. K. Omura. *Principles of Digital Communication and Coding.* McGraw-Hill Book, New York, 1979.

[80] B. Widrow. A study of rough amplitude quantization by means of Nyquist sampling theory. *IRE Trans. Circuit Theory*, CT-3:266–276, 1956.

[81] B. Widrow. Statistical analysis of amplitude quantized sampled data systems. *Trans. Amer. Inst. Elec. Eng., Pt. II: Applications and Industry*, 79:555–568, 1960.

[82] P. W. Wong and R. M. Gray. Two stage sigma-delta modulation. *IEEE Transactions on ASSP*, 1989. To Appear.

[83] A. Wyner and J. Ziv. Bounds on the rate-distortion function for stationary sources with memory. *IEEE Transactions on Information Theory*, 17:508–513, Sept 1971.

[84] Y. Yamada, S. Tazaki, and R. M. Gray. Asymptotic performance of block quantizers with a difference distortion measure. *IEEE Transactions on Information Theory*, IT-26:6–14, Jan. 1980.

[85] P. L. Zador. *Development and evaluation of procedures for quantizing multivariate distributions.* Ph. D. dissertation, Stanford University, 1963.

[86] P. L. Zador. Topics in the asymptotic quantization of continuous random variables, 1966. Unpublished Bell Laboratories Memorandum.

[87] P. L. Zador. Asymptotic quantization error of continuous signals and the quantization dimension. *IEEE Transactions on Information Theory*, IT-28:139–148, Mar. 1982.

[88] J. Ziv. On universal quantization. *IEEE Transactions on Information Theory*, IT-31:344–347, 1985.

Index

9 780792 390480